Anesthesia Care of Patients with Skeletal Dysplasia

Mary C. Theroux
Editor

Anesthesia Care of Patients with Skeletal Dysplasia

Springer

Editor
Mary C. Theroux, MD
Department of Anesthesiology
and Perioperative Medicine
Nemours Children's Hospital
Wilminton, Delaware, USA

Professor of Anesthesiology and Pediatrics,
Sidney Kimmel Medical College
Thomas Jefferson University
Philadelphia, Pennsylvania, USA

ISBN 978-3-031-92500-9 ISBN 978-3-031-92501-6 (eBook)
https://doi.org/10.1007/978-3-031-92501-6

This Springer imprint is published by the registered company Springer Nature Switzerland AG
The registered company address is: Gewerbestrasse 11, 6330 Cham, Switzerland

To all the patients with skeletal dysplasia, your strength, courage, and resilience have inspired every page of this book. Your stories illuminate the path of understanding and compassion in medicine.

And to my late husband, whose tremendous support, unwavering belief, and undying love made this work possible. Your spirit continues to guide me every step of the way.

This book is for you.

Preface

In *Anesthesia Care of Patients with Skeletal Dysplasia*, medical science and patient care merge together as we discuss the multitude of anesthetic complexities for people with skeletal dysplasia. I have spent my career as an anesthesiologist and physician-researcher at Nemours Children's Hospital in the Delaware Valley (formerly known as Alfred I. duPont Hospital for Children), a premier center for skeletal dysplasia care. My career has focused on the safety of anesthesia for patients with skeletal dysplasia. This book bridges the knowledge gap between general anesthesia providers and specialists at Nemours, who are uniquely experienced in managing these complex cases. It is my honor to share this experience with you.

Historically, skeletal dysplasia has been classified or categorized based on clinical and radiological findings. With advancement in the field of genetics, this classification has evolved into a "nosology of genetic skeletal disorders." Since its inception, this nosology has undergone 11 revisions, with the most current version published in 2023 [1]. In this book, the authors refer to nosology only to the extent that it aids in understanding similarities between phenotypes and, therefore (to some degree), the similarities in anesthetic risks between phenotypes. We strive to provide a similar grouping as mentioned in the nosology of 2023; however, we have also tried to maintain the identity and nomenclature of some of the dysplasia(s) as they may be known to an anesthesiologist in the community. Thus, when we do not strictly adhere to the current nosology, we seek the reader's understanding of the rationale. Because this book was written for all anesthesia practitioners, most of whom might only have a superficial knowledge of genetics, the content is tailored to the expectations of anesthesia providers both in the community and in university centers.

This book is intentionally redundant in the descriptions of the anesthetic management of dysplasia; each chapter is comprehensive and self-sufficient, allowing the readers to fully understand the care required for a specific patient without cross-referencing other chapters. Not only does this book discuss the management modalities of anesthesia, but it also emphasizes the importance of a multidisciplinary approach to the overall care and success of the perioperative management of these complex patients.

A crucial aspect of this book is its emphasis on the patient's perspective. Living with skeletal dysplasia involves unique challenges and triumphs. It is essential to recognize the resilience and diversity of experiences within this community, especially during medical care. The authors include personal stories and insights from individuals with skeletal dysplasia and the difficulties they have experienced while navigating and advocating for their own anesthesia care. The landscape of skeletal dysplasia is continually evolving, and it is our hope that this book contributes to that evolution by informing, inspiring, and igniting a dialogue among all anesthesia caregivers.

This work is a resource primarily for anesthesia providers and their patients with skeletal dysplasia who are seeking to improve anesthesia care, enhance safety, and minimize complications. It reflects my gratitude to those who have shared their experiences and knowledge to help achieve this goal.

All contributing authors of this book are from Nemours Children's Hospital, Delaware Valley. It is common for medical books to feature authors from various institutions, but our team has unique expertise and experience managing patients with skeletal dysplasia, a population less familiar to most other institutions. The book also highlights the specialized knowledge required for these cases and underscores the proficiency and capability of our anesthesia staff in providing safe care for patients with skeletal dysplasia.

In closing, I extend my gratitude to the patients who have shared their experiences and knowledge which make this book a comprehensive resource not only for anesthesia providers but also for patients with skeletal dysplasia.

Philadelphia, Pennsylvania, USA Mary C. Theroux

Reference

1. Unger S, Ferreira CR, Mortier GR, Ali H, Bertola DR, Calder A, et al. Nosology of genetic skeletal disorders: 2023 revision. Am J Med Genet A. 2023;191:1164–209.

Acknowledgment

This book is the culmination of 5 years of dedication, research, and collaboration. I extend my deepest gratitude to those whose support and expertise have been invaluable throughout this journey.

I am especially grateful to **Dr. Freeman Miller** for his mentorship in medical writing, which helped shape this book. His guidance, insights, and unwavering support throughout the writing process have been instrumental in shaping this book.

My sincere appreciation also goes to **Drs. William Mackenzie**, **Stuart Mackenzie**, and **Jeffrey Campbell** for their clinical insights and support reinforcing the importance of multidisciplinary care in the skeletal dysplasia population.

To my colleagues in the Department of Anesthesiology and Perioperative Medicine, your expertise, discussions, and commitment to patient care have enriched this book. I wish to recognize **Dr. Vanessa Olbrecht** for her enthusiasm, tireless energy, and unwavering passion for advancing anesthesia care. Her encouragement and commitment have been both motivating and deeply appreciated.

I also extend my gratitude to the Departments of Orthopedics, Neurosurgery, Orthogenetics, Radiology, Otolaryngology, and Pulmonary Medicine. Their collaboration and shared knowledge have been instrumental in highlighting the multidisciplinary approach essential to caring for patients with skeletal dysplasia.

Lastly, I am deeply thankful to the patients and families who have entrusted us with their care. Their resilience, courage, and experiences have been the driving force behind this work, reminding us of the profound impact of compassionate and well-informed medical care.

To my family, friends, and all those who have supported me through this journey, I extend my heartfelt thanks. Your patience, understanding, and encouragement have made this book possible. It is my sincere hope that this book will serve as a valuable resource for advancing anesthesia care for patients with skeletal dysplasia.

2-17-2025 Mary C. Theroux

Contents

Editor and Contributors

About the Editor

Mary C. Theroux, MD With over 35 years of distinguished experience as an anesthesiologist at Nemours Children's Health in Wilmington, Delaware, USA—an institution internationally recognized for excellence in managing patients with skeletal dysplasia—the author provides an exceptional depth of expertise in the specialized anesthetic care required for these patients. This book aims to enhance clinical practice and patient outcomes by presenting expert insights from a leader devoted to advancing the anesthetic care of patients with skeletal dysplasia.

Contributors

Lynda Arai, MD Department of Anesthesiology and Perioperative Medicine, Nemours Children's Hospital, Wilmington, Delaware, USA
Anesthesiology and Pediatrics, Sidney Kimmel Medical College, Thomas Jefferson University, Philadelphia, Pennsylvania, USA

Lauren W. Averill, MD Radiology and Pediatrics, Sidney Kimmel Medical College, Thomas Jefferson University, Philadelphia, Pennsylvania, USA
Department of Radiology, Nemours Children's Hospital, Wilmington, Delaware, USA

Melinda Brown, MS Department of Anesthesiology and Perioperative Medicine, Nemours Children's Hospital, Wilmington, Delaware, USA

Janine Catanach, CPNP Department of Anesthesiology and Perioperative Medicine, Nemours Children's Hospital, Wilmington, Delaware, USA

Sabina Dicindio, DO Anesthesiology and Pediatrics, Sidney Kimmel Medical College, Thomas Jefferson University, Philadelphia, Pennsylvania, USA
Department of Anesthesiology and Perioperative Medicine, Nemours Children's Hospital, Wilmington, Delaware, USA

Anthony DiNardo, DC Department of Specialty Care, Nemours Children's Hospital, Wilmington, Delaware, USA

Colleen P. Ditro, DNP Clinical Operations, Orthopedics, Nemours Children's Hospital, Wilmington, Delaware, USA

Divya Dixit, MD Department of Anesthesiology and Perioperative Medicine, Nemours Children's Hospital, Wilmington, Delaware, USA
Anesthesiology and Pediatrics, Sidney Kimmel Medical College, Thomas Jefferson University, Philadelphia, Pennsylvania, USA

Alfred T. Dorsey, MD Department of Anesthesiology and Perioperative Medicine, Nemours Children's Hospital, Wilmington, Delaware, USA
Anesthesiology and Pediatrics, Sidney Kimmel Medical College, Thomas Jefferson University, Philadelphia, Pennsylvania, USA

Phoebe Fisher, MD Anesthesiology and Pediatrics, Sidney Kimmel Medical College, Thomas Jefferson University, Philadelphia, Pennsylvania, USA
Department of Anesthesiology and Perioperative Medicine, Nemours Children's Hospital, Wilmington, Delaware, USA

Megan Grannan, BA Williamstown, New Jersey, USA

Judith Jones, MD Department of Anesthesiology and Perioperative Medicine, Nemours Children's Hospital, Wilmington, Delaware, USA
Anesthesiology and Pediatrics, Sidney Kimmel Medical College, Thomas Jefferson University, Philadelphia, Pennsylvania, USA

Robert S. Lang, MD Department of Anesthesiology and Perioperative Medicine, Nemours Children's Hospital, Wilmington, Delaware, USA
Anesthesiology and Pediatrics, Sidney Kimmel Medical College, Thomas Jefferson University, Philadelphia, PA, USA

Doyle Lim, MD, MMM Department of Anesthesiology and Perioperative Medicine, Nemours Children's Hospital, Wilmington, Delaware, USA
Anesthesiology and Pediatrics, Sidney Kimmel Medical College, Thomas Jefferson University, Philadelphia, Pennsylvania, USA

Connie J. Lin, MD Department of Anesthesiology and Perioperative Medicine, Nemours Children's Hospital, Wilmington, Delaware, USA
Anesthesiology and Pediatrics, Sidney Kimmel Medical College, Thomas Jefferson University, Philadelphia, Pennsylvania, USA

Sharon Luo, MD Department of Anesthesiology and Perioperative Medicine, Nemours Children's Hospital, Wilmington, Delaware, USA
Anesthesiology and Pediatrics, Sidney Kimmel Medical College, Thomas Jefferson University, Philadelphia, Pennsylvania, USA

Malgorzata Lutwin-Kawalec, MD Department of Anesthesiology and Perioperative Medicine, Nemours Children's Hospital, Wilmington, Delaware, USA
Anesthesiology and Pediatrics, Sidney Kimmel Medical College, Thomas Jefferson University, Philadelphia, Pennsylvania, USA

Matthew McCoy, DNP Department of Anesthesiology and Perioperative Medicine, Nemours Children's Hospital, Wilmington, Delaware, USA

Susmita Nesargi, MD Department of Anesthesiology and Perioperative Medicine, Nemours Children's Hospital, Wilmington, Delaware, USA
Anesthesiology and Pediatrics, Sidney Kimmel Medical College, Thomas Jefferson University, Philadelphia, Pennsylvania, USA

Vanessa A. Olbrecht, MD, MBA Anesthesiology and Pediatrics, Sidney Kimmel Medical College, Thomas Jefferson University, Philadelphia, Pennsylvania, USA
Department of Anesthesiology and Perioperative Medicine, Nemours Children's Hospital, Wilmington, Delaware, USA

Robert F. Olszewski, MD Clinical Anesthesiology, Lewis Katz School of Medicine, Temple University, Philadelphia, Pennsylvania, USA

Abraham Oommen, MD Department of Anesthesiology and Perioperative Medicine, Nemours Children's Hospital, Wilmington, Delaware, USA
Anesthesiology and Pediatrics, Sidney Kimmel Medical College, Thomas Jefferson University, Philadelphia, Pennsylvania, USA

Nathalie Peiris, MD Department of Anesthesiology and Perioperative Medicine, Nemours Children's Hospital, Wilmington, Delaware, USA
Anesthesiology and Pediatrics, Sidney Kimmel Medical College, Thomas Jefferson University, Philadelphia, Pennsylvania, USA

Joseph Piatt, MD Department of Neurosurgery, Nemours Children's Hospital, Wilmington, Delaware, USA

Manish Purohit, MD Anesthesiology and Pediatrics, Sidney Kimmel Medical College, Thomas Jefferson University, Philadelphia, Pennsylvania, USA
Department of Anesthesiology and Perioperative Medicine, Nemours Children's Hospital, Wilmington, Delaware, USA

Isabel Rosenberg, CPNP Department of Anesthesiology and Perioperative Medicine, Nemours Children's Hospital, Wilmington, Delaware, USA

Carly Ruhnke Little but Tough, Toms River, New Jersey, USA

Kesavan Sadacharam, MD Department of Anesthesiology and Perioperative Medicine, Nemours Children's Hospital, Wilmington, Delaware, USA
Anesthesiology and Pediatrics, Sidney Kimmel Medical College, Thomas Jefferson University, Philadelphia, Pennsylvania, USA

Eric Stickles, MD Department of Anesthesiology and Perioperative Medicine, Nemours Children's Hospital, Wilmington, Delaware, USA

Mary C. Theroux, MD Department of Anesthesiology and Perioperative Medicine, Nemours Children's Hospital, Wilminton, Delaware, USA
Professor of Anesthesiology and Pediatrics, Sidney Kimmel Medical College, Thomas Jefferson University, Philadelphia, Pennsylvania, USA

Sheaba L. Varghese, MD Anesthesiology and Pediatrics, University of Cincinnati, Cincinnati, Ohio, USA
Cincinnati Children's Hospital Medical Center, Cincinnati, Ohio, USA

Part I

Organizing Care for Patients with Skeletal Dysplasia

Introduction and Overview

1

Lynda Arai, Vanessa A. Olbrecht, and Mary C. Theroux

Skeletal dysplasias are a heterogeneous group of inheritable disorders that impair growth in children and affect multiple organ systems. These conditions present in infancy and require lifelong treatment for a variety of conditions. As a result, patients with skeletal dysplasia see multiple subspecialists, and finding both appropriate and comprehensive care is often challenging. Some skeletal dysplasias are relatively common and easier to manage; others are exceedingly rare. Finding appropriate specialists with the necessary expertise in managing the disease and associated comorbidities can take significant time and effort. Therefore, it is crucial to develop a team of specialists within a tertiary care center who can address all needs of patients with skeletal dysplasia.

Because skeletal dysplasias are inherited disorders, genetics play a pivotal role in advancing and understanding of these disorders. Consequently, geneticists play an integral role in caring for patients with skeletal dysplasias. They often serve as the primary care physicians for the patients, guiding diagnosis, finding appropriate subspecialty care, and providing long-term management. The 2023 revision of the nosology of genetic skeletal disorders [1], now in its 11th update, marks significant advancements in classifying and understanding these disorders. The revision lists a total of 771 disorders associated with 552 genes and categorized into 41 groups based on clinical, radiographic, and molecular phenotypes. This nosology is expected to expand with the inclusion of newly identified skeletal dysplasias and the

L. Arai (✉) · V. A. Olbrecht · M. C. Theroux
Department of Anesthesiology and Perioperative Medicine, Nemours Children's Hospital, Wilmington, Delaware, USA

Anesthesiology and Pediatrics, Sidney Kimmel Medical College, Thomas Jefferson University, Philadelphia, Pennsylvania, USA
e-mail: larai@nemours.org; Vanessa.Olbrecht@nemours.org; mtheroux@nemours.org

M. C. Theroux (ed.), *Anesthesia Care of Patients with Skeletal Dysplasia*,
https://doi.org/10.1007/978-3-031-92501-6_1

further sub-classification of existing dysplasias. A detailed description of the current nosology, however, is beyond the scope of this book and may not necessarily enhance the specific knowledge required by anesthesia providers.

Most patients with skeletal dysplasia will require at least one surgical intervention during their life, not only to manage their skeletal abnormalities, but also to treat other affected organ systems. Unfortunately, many institutions are unfamiliar with the anesthetic implications associated with each disorder, and the literature is sparse. Thus, more published information is needed to guide anesthesia providers when caring for patients with these disorders. Anesthetic care is a critical aspect of the safe management of these patients in the perioperative period, and lack of knowledge or familiarity with an individual disorder can lead to dire consequences. While patients often recognize the importance of selecting a skilled surgeon, many may not fully appreciate the necessity of an equally competent anesthesia team. Anesthesia schedules present further challenges. Assignments are scheduled only 1–2 days in advance, which can result in the random placement of anesthesiologists to care for these complex patients. This emphasizes the importance of a skeletal dysplasia team to aid in the early identification of surgical patients and to coordinate with the anesthesia department to ensure the availability and assignment of a suitable anesthesiologist on the day of surgery. While this level of coordination can work well for elective procedures, it does not help with urgent and emergent procedures. Often, schedules must be adjusted to ensure the most appropriate care. Furthermore, anesthesiologists may need additional support from colleagues or from an otolaryngologist familiar with the unique airway challenges of patients with skeletal dysplasia, especially if significant airway management difficulties are anticipated—a common issue in skeletal dysplasia disorders.

Our institution has established a skeletal dysplasia team comprising specialists with extensive knowledge and expertise in caring for children with various forms of skeletal dysplasia. This group, developed over several years, includes experts from orthopedic surgery, neurosurgery, anesthesiology, genetics, radiology, and otolaryngology, as well as advanced practice nurse specialists dedicated to this field. Other essential team members include physiotherapists, occupational therapists, and social workers. Much of our expertise has come from experience and collaboration with other specialists in their respective fields at institutions worldwide. The publication of information on the care of patients with specific skeletal dysplasias has been critical in advancing knowledge and improving care for these disorders. The dedication and tenacity of a small group of individuals recognized and supported by leadership and administration have been pivotal in providing the best possible anesthesia care to these medically challenging patients. Open communication with the surgical team is an integral component of the structure and success of our team. When caring for a patient with skeletal dysplasia, we maintain a healthy caution, aware that these patients can experience life-threatening complications at any point during the perioperative period.

1.1 Overview

Part I of the book provides insight into the challenges an anesthesiology department faces when caring for complex patient populations, such as those with skeletal dysplasia. It highlights the negative impact of fragmented care, wherein multiple providers unfamiliar with a patient's unique anesthetic challenges are involved. This section is further enhanced by narratives from the patients themselves. Emphasizing a multidisciplinary approach to care, Part I details how such care is possible within the framework of a tertiary health care system.

Part II discusses the anesthetic management of commonly seen individual skeletal dysplasias and includes a description of the relevant characteristics of each one. As discussed in the book's preface, genetic backgrounds are mentioned only briefly.

Part III discusses the use of neuromonitoring, which has become increasingly valued in assessing and ensuring spinal cord integrity for dysplasias that carry a risk of spinal injury (i.e., "spine at risk"). This section also includes a chapter specifically dedicated to the anesthetic management of procedures involving the spine.

Part IV discusses pain management using conventional methods as well as peripheral nerve blocks, which have been used increasingly in anesthesia. Many patients with skeletal dysplasia are unsuitable for neuraxial anesthetics, so peripheral blocks have come to play an important role in optimizing analgesia while minimizing the use of opioids.

Patients with skeletal dysplasia may have multiple medical issues and concerns related to their skeletal disorder and other affected organ systems. They should seek a care center with specialists who have experience and expertise in a wide range of skeletal dysplasias. It is equally important that the specialists have a collaborative approach to managing these challenging patients to ensure they receive the best care possible while minimizing the risk of adverse outcomes.

Reference

1. Unger S, Ferreira CR, Mortier GR, et al. Nosology of genetic skeletal disorders: 2023 revision. Am J Med Genet A. 2023;191:1164–209.

Navigating an Anesthesiology Department and Finding the Care You Need: Patient Perspective

2

Megan Grannan, Carly Ruhnke, Isabel Rosenberg, and Janine Catanach

2.1 Patient 1

2.1.1 Introduction

My name is Megan, and I am a 31-year-old woman living with metatropic dysplasia, a form of skeletal dysplasia. Throughout my life, I have undergone countless surgeries and faced numerous challenges related to anesthesia. The skeletal dysplasia population faces unique surgical and anesthetic complications, and, in my experience, many health care providers lack the necessary experience to care for patients like me. They often do not fully understand my complex medical and psychosocial issues that result from multiple complex surgeries, nor do they appreciate the impact my condition has had on my family.

2.1.2 Early Experiences with Medical Care

One of my earliest recollections of medical care was at a local hospital, where several anesthesiologists expressed concern that my airway was "too small." They were worried that my small oral opening and facial features might make intubation (the

M. Grannan
Williamstown, New Jersey, USA

C. Ruhnke
Little but Tough, Toms River, New Jersey, USA

I. Rosenberg · J. Catanach (✉)
Department of Anesthesiology and Perioperative Medicine, Nemours Children's Hospital, Wilmington, Delaware, USA
e-mail: Isabel.rosenberg@nemours.org; Janine.catanach@nemours.org

M. C. Theroux (ed.), *Anesthesia Care of Patients with Skeletal Dysplasia*,
https://doi.org/10.1007/978-3-031-92501-6_2

placement of a breathing tube) more difficult. This led to a referral to a facility with more experience in caring for patients with skeletal dysplasia. Although my family and I had limited medical knowledge at the time, I am now incredibly thankful for those early anesthesiologists' awareness and recognition of my difficult airway.

When we began receiving care at a more experienced children's hospital, we encountered a proficient and collaborative team. This team included specialists in anesthesia, orthopedics, and otolaryngology (ENT; ear, nose, and throat) as well as experienced nurses who knew how to communicate with me and help me through my anxieties. Witnessing this level of collaboration among medical professionals was an eye-opening experience, and knowing that the team had my safety in mind was immensely comforting.

2.1.3 Challenges in Ambulatory Surgical Centers

As the years went on, some of my procedures were attempted at ambulatory surgical centers due to the nature of the surgery—in one instance, a dental procedure. However, some of these experiences were disastrous. I vividly recall one particular instance where a "simple" tooth extraction ended with me on a ventilator due to multiple failed intubation attempts. I woke up with a tube down my throat, terrified of this unexpected outcome. Equally distressing was seeing the doubt and fear in my parents' eyes. We all learned, perhaps in the hard way, that a hospital setting was the safest place for me to have surgery.

2.1.4 Airway Evaluation and Traumatic Experiences

Eventually, I underwent an airway evaluation with an otolaryngologist, commonly known as an ENT specialist. Using a camera and numbing cream to suppress my gag reflex, they visualized my airway and determined that it was shaped like a hockey stick. Despite this clear abnormality, I still encountered overconfident anesthesiologists who believed they could easily manage my airway. Unfortunately, one such anesthesiologist attempted to intubate me blindly 17 times without calling for support from other specialists. This experience left me in the ICU with severe throat and tongue swelling and a permanent traumatic memory.

On the other hand, I have also encountered anesthesiologists who admit they are uncomfortable with my airway. Hearing statements like "I've never seen anyone like you" has been traumatic in a different, more emotional way. These moments have reinforced the importance of having a medical team that truly understands the complexities of skeletal dysplasia.

2.1.5 The Emotional Impact of My Surgical Experiences

Beyond airway issues, there are many other challenges to consider. Among these are unanticipated difficulties with IV access and the emotional toll that multiple complex surgeries can have on both the patient and their family. Due to the nature of my skeletal dysplasia, I have undergone some of the longest and most difficult surgeries, including hip replacements and multiple spinal procedures. Enduring these complex surgeries and their prolonged recoveries has undoubtedly made me a stronger person, but it has also taken a significant toll on my emotional health. Through the years, I have watched my family cope with many terrifying surgical experiences, putting on a brave face despite the fear of the unknown. Although these issues may not be exclusive to the skeletal dysplasia population, they are important to bring awareness to. I am personally thankful for the nurses, nurse practitioners, and social workers who have helped me through these often-overlooked concerns.

2.1.6 My Adult Perspective Regarding My Specialized Anesthesia Care Needs

As I grew into adulthood, my airway difficulties progressively worsened, a feature common to patients with skeletal dysplasia. When I have a scheduled procedure, my main fear is related to the anesthesia team's ability to manage my airway without causing harm. Having an anesthesiologist and a medical team experienced with dwarfism is almost more important to me than the surgery itself.

I have been fortunate to develop a strong connection with my anesthesiologist, Dr. Mary C. Theroux, the editor of this book and a well-known expert in the field of skeletal dysplasia. She knows when I am stressed and how to help me through my most fearful moments, especially in those final moments before I am wheeled into the operating room. In fact, I keep Dr. Theroux's number marked in my phone as an emergency contact. This might seem excessive, but I have had too many terrible experiences at community hospitals with doctors who overconfidently assume they know how to care for a patient with skeletal dysplasia. My bad experiences have taught me to advocate for myself. Now, as an adult, I insist that my care team is in touch with Dr. Theroux and that multiple anesthesiologists are available in the event of a problem.

2.1.7 My Message to Anesthesiologists

To the anesthesiologists reading this, I invite you to learn from my experiences. Be prepared. Be confident, but stay humble. Educate your patient, and do not forget that the family is your patient as well. Most importantly, be yourself and be there for your patient. Earning the trust of your patient is of the utmost importance.

2.2 Patient 2

2.2.1 Introduction

My name is Carly. I am a 28-year-old woman with Morquio syndrome type A. My medical journey started out with many years of uncertainty about my condition, but, luckily, I found my way to a medical team that provides me with the specialized care that I need.

2.2.2 Early Life and Diagnosis

My story begins even before my birth, when my mother's intuition alerted her that something wasn't quite right during her pregnancy. Despite advanced ultrasounds revealing multiple physical abnormalities, nothing conclusive was diagnosed. When I was born, I had club feet, multiple hemangiomas, a clicking hip, enlarged organs, wide-set eyes, and a large head. Despite these physical differences, no official diagnosis was made. At 6 months old, I was diagnosed with kyphoscoliosis, yet there was still no awareness of my skeletal dysplasia. My parents endured multiple missed diagnoses before I was ultimately diagnosed with Morquio syndrome. The adjustment to this catastrophic diagnosis was grueling, but once we moved past the gravity of it, my parents became committed to finding the best possible medical care.

2.2.3 Finding Specialized Care

Our search for doctors with expertise in skeletal dysplasia led us to Dr. Steven Kopits. The first surgery he performed was a cervical and thoracic spinal fusion to correct spinal cord impingement. This 16-h surgery required a 6-month recovery in a halo and a body cast. It was through these early surgical experiences that we first learned how challenging anesthesia would be for me. Intubating my airway was difficult due to limited neck mobility, a small oral opening, and the unique shape and size of my airway. This turned out to be a recurring issue and has been the most difficult part of my anesthesia care.

2.2.4 Traumatic Surgical Experiences

One of my most traumatic surgical memories was during an ear tube placement procedure. Despite our requests for a specialized anesthesiologist, we were reassured it wasn't necessary since it was a "minor" procedure. Unfortunately, this overconfident anesthesiologist was unable to manage my airway, which led me to have a near-respiratory-arrest situation. An emergency backup team rescued my airway. What was supposed to be a quick procedure turned into a 4-h anesthesia and surgery, and the trauma of that experience lives with me to this day.

2.2.5 Finding the Right Medical Team

Eventually, I ended up in the care of Dr. William Mackenzie, an orthopedic surgeon with expertise in all forms of skeletal dysplasia, and Dr. Mary Theroux, an esteemed pediatric anesthesiologist skilled in difficult airways and in the care of patients with skeletal dysplasia. Whenever I need surgery, I insist that Dr. Theroux is involved. She has developed strategies to keep me safe, including placing a detailed note in my medical record about the intricacies of my airway and recommendations for successful intubation. She determined it was safer for me to be awake during intubations with the help of local anesthetics and mild sedatives. She is even my first emergency contact in my cell phone!

2.2.6 Conclusion

I am incredibly grateful for the wonderful doctors at Nemours who have the knowledge and expertise to keep me safe. Patients with skeletal dysplasia inherently have many risks related to their disease, and the risk of difficult airway and intubation is common with many forms of skeletal dysplasia. Families of children with Morquio syndrome must feel empowered to voice their concerns, not only about surgery, but also about anesthesia care, and it is imperative that their concerns are acknowledged and addressed by a proficient anesthesia team.

Support and Management of the Skeletal Dysplasia Surgical Program: Challenges Facing an Anesthesiology Department

3

Doyle Lim, Lynda Arai, Janine Catanach, and Robert F. Olszewski

This chapter has two main focuses:

1. Provide an overview of anesthesia care to inform and educate our patients.
2. Describe the operational difficulties of an anesthesiology department, specifically regarding "hands-on" care of patients, which may be an area unfamiliar to the patients.

D. Lim (✉) · L. Arai
Department of Anesthesiology and Perioperative Medicine, Nemours Children's Hospital, Wilmington, Delaware, USA

Anesthesiology and Pediatrics, Sidney Kimmel Medical College, Thomas Jefferson University, Philadelphia, Pennsylvania, USA
e-mail: doyle.lim@nemours.org; larai@nemours.org

J. Catanach
Department of Anesthesiology and Perioperative Medicine, Nemours Children's Hospital, Wilmington, Delaware, USA
e-mail: Janine.catanach@nemours.org

R. F. Olszewski
Clinical Anesthesiology, Lewis Katz School of Medicine, Temple University, Philadelphia, Pennsylvania, USA
e-mail: robert.olszewski@tuhs.temple.edu

M. C. Theroux (ed.), *Anesthesia Care of Patients with Skeletal Dysplasia*,
https://doi.org/10.1007/978-3-031-92501-6_3

3.1 Overview of Anesthesia Care

There are several types and levels of anesthesia, starting with local anesthetics that numb specific areas; to mild, moderate, or deep sedation, which reduces consciousness and response to painful stimuli; and finally, general anesthesia, which renders a patient unconscious while the anesthesia team supports vital functions to provide a safe and stable anesthetic depth. While sedation can be employed for patients undergoing minor procedures or medical imaging, general anesthesia is often required for procedures in the operating room (OR). General anesthesia may be described as a "controlled unconsciousness during which the patient is unaware of surgery and does not move or feel pain to a painful stimulus." General anesthesia can be combined with regional anesthesia techniques, such as nerve blocks, to provide multiple pathways for intra-operative and postoperative pain management. General anesthesia often requires controlling the patient's breathing using "airways," such as an endotracheal tube (breathing tube) or a laryngeal mask airway (LMA). The method used to accomplish this often is termed "airway management" and is a key point in caring for patients with SD.

In addition to airway management, intravenous (IV) access is necessary to administer fluids and provide anesthetic agents or other medications to the patient should the need arise. The anesthesiology team also monitors and manages other potential concerns, such as cardiac, pulmonary, and/or neurologic conditions, thus ensuring the patient's care is tailored to their individual health needs.

3.2 Timeline to Surgery/Patient Evaluation

The first contact for a skeletal dysplasia patient with a perioperative program usually comes through the patient's primary care doctor. Before any surgical procedure, the patient's medical history is reviewed, followed by a discussion about the available types of anesthesia and the risks and alternatives to each. A physical exam and an airway assessment are performed on the day of surgery. This comprehensive assessment enables the anesthesiologist to develop an individualized anesthetic plan. Anesthesiologists and preoperative advanced practice nurses lead this process. Information gathered includes prenatal history, past and present medical conditions (including severity and treatments), recent assessments by specialty services, home medications, any necessary respiratory support, and recent imaging studies. Records from any previous anesthetics are especially helpful if available. Additional considerations covered include home care needs, travel, housing, and financial assistance (if necessary). E-mail, video, or phone calls, along with secure mobile apps, facilitate communication and coordination, streamline the preoperative process, and ensure all necessary information is gathered efficiently.

3.2.1 Week(s) Before Surgery

As the surgery date approaches, many programs utilize checklists to ensure that all necessary steps are completed. These typically include updated radiographic and laboratory evaluations, which help assess the patient's condition before surgery. Ideally, the patient should be in the best possible health prior to surgery, which has been shown to reduce postoperative complications, shorten the hospital stay, and improve recovery and overall patient experience.

At our institution, discussions of upcoming patient encounters and needs are held weekly by a multi-disciplinary skeletal dysplasia team made up of a neurosurgeon, orthopedists, anesthesiologists, geneticists, otolaryngologists, hospitalists, pulmonologists, and other specialists as needed. These discussions focus on the unique aspects of each patient's surgical and anesthesia plan, identifying areas that may need further investigation or additional information. Recommendations to optimize the patient's health, such as adjusting medications, conducting further tests, or addressing nutritional needs, are made during these meetings. Resources for psychological and family support are also offered to assist patients throughout the process.

Travel to a tertiary or quaternary center may require arranging accommodations for the family and patient prior to surgery. Our institution works with our local Ronald McDonald House (located on-site) and offers options like recreational vehicle parking and support hookups to ensure a comfortable stay before surgery.

3.2.2 Day of Surgery and Recovery

On the day of surgery, after registration, the patient is brought to a preoperative holding area where their medical history is updated and informed consent for surgery and anesthesia is obtained. This is an opportunity for patients and families to ask any last-minute questions and set realistic expectations. Pre-medications to reduce anxiety and/or pain are typically given at this time. After checklists are reviewed and finalized, the patient is brought to the operating room.

Following surgery, the patient will be cared for in a regular unit/ward or an intensive care unit (if necessary). A standardized handoff facilitates the transfer of patient care to the inpatient or critical care team. The orthopedics, pain management, respiratory therapy, and physical and occupational therapy teams will also follow the patient to address individual recovery needs. Patient Education will prepare the patient and family for continued recovery and rehabilitation after discharge, and Social Work will facilitate support for home care and disposition planning.

3.2.3 Day of Surgery Timeline

On the day of surgery, the patient will be asked to arrive at the hospital 1–2 h in advance to facilitate check-in and final assessment and paperwork. During this time, the patient will meet their anesthesia team and the nurses who will be assisting in their care. Intravenous access (IV) may be obtained in the holding area during this time. The patient will also likely be offered medications to help ease anxiety and to assist in postoperative pain control. When the operating room is ready, the anesthesia team will transport the patient to the operating room on a stretcher.

In the operating room, the anesthesia team will apply monitors to closely monitor heart rate and rhythm, oxygen saturation of blood (pulse oximeter), and blood pressure. If IV access has been obtained in the preoperative area, a mask may be placed with oxygen over the patient's mouth and nose as a safety measure. Medications will then be administered via the IV to induce sleep. If an IV was not obtained in the preoperative area, the anesthesia team may perform an "inhalational induction" (of anesthesia) using a mask to deliver medications to induce sleep. In this situation, the anesthetics are absorbed via the lungs. After a sufficient level of anesthesia is achieved, IV access is then obtained. It is important to note that in cases where a difficult airway or presence of certain coexisting conditions ("comorbidities") are present, it is safer to obtain IV access prior to the administration of anesthesia. After induction of anesthesia, the care team will place a device (a breathing tube or a laryngeal mask airway) to help the patient breathe during the surgery, with the expectation that it will be removed at the end of the procedure. In most cases, this device will be placed after the patient is asleep; however, in some circumstances, it may need to be placed before induction of anesthesia while the patient is still awake. In these rare cases, the anesthesiologist will discuss this with the patient beforehand and will offer some sedation and numbing medicine to make the experience less unpleasant.

Most patients will not remember anything once anesthesia takes effect, and their next memory will be waking up in the recovery room. There, monitors similar to those used in the operating room will be reapplied, and oxygen will be delivered through a mask until the patient is fully awake. Medications will be available to treat any pain or nausea that the patient may experience. Depending on the complexity of the surgery, the patient will either be transferred to an intensive care unit, inpatient floor or discharged home.

3.2.4 Emotional Preparation for Surgery

Surgery and anesthesia can make even experienced patients nervous. This effect is compounded in children whose cognitive formation is still ongoing. Excessive anxiety in the patient can negatively impact the patient's experience and recovery. Anxiety increases the perception of pain and can lead to complications on emergence from anesthesia, such as delirium. Therefore, it is essential to implement strategies to mitigate preoperative and day of surgery anxiety in all patients. Preparing the child by discussing what to expect, encouraging questions, and addressing fears can help reduce anxiety. Familiarizing the child with the hospital environment through books, videos, or preoperative tours can alleviate fear. Comfort items such as a favorite toy or blanket may be brought to the hospital to provide emotional comfort on the day of surgery.

Parents should also remember to care for their own emotional well-being. Children can sense when their parents are anxious, so staying calm and positive can help the child feel more secure.

3.3 Operational Difficulties Unique to Anesthesiology

Unlike many medical specialties that foster long-term relationships between patients and their physicians, anesthesiology departments typically operate under a team-based model. In this model, patients are assigned to an anesthesia provider based on availability rather than any prior relationship. This team-based approach within anesthesiology has both benefits and drawbacks. The most notable advantage is its flexibility in ensuring that an anesthesiologist is always available for all patients scheduled for surgery on any given day. This flexibility is necessary since the patients primarily are scheduled based on surgeon availability, and it eliminates the challenges of coordinating multiple schedules to pair a specific anesthesiologist and a surgeon with a particular patient. The main drawback of this model is the loss of comfort and familiarity that comes from repeatedly having the same anesthesiologist care for a patient. This aspect is particularly critical in complex medical cases involving patients with skeletal dysplasia, where continuity of care can significantly impact safety. Consequently, there are circumstances in which it is crucial to maintain the continuity of care by ensuring a consistent patient–anesthesiologist relationship.

At our institution, special care is taken to ensure that patients with skeletal dysplasia are matched with anesthesiologists experienced in managing the unique challenges posed by their condition. If surgery is scheduled at a facility without experience in skeletal dysplasia, it is important to contact the anesthesiology department in advance to inform them of the patient's diagnosis and potential challenges, particularly regarding airway management. Documentation from previous anesthetic experiences, including any letters regarding difficult airway management, can be extremely helpful in formulating a safe and effective care plan.

Navigating the process of contacting the anesthesiology department can be daunting. Surgeons and their office staff often know the anesthesiologists personally and can help facilitate contact. If the surgery is being performed at a hospital, inquire about a "preoperative optimization clinic," which is often run by the anesthesiology department and serves as an effective point of contact. Alternatively, preoperative nurses may be able to flag complex cases for anesthesiologist follow-up. For office-based procedures, such as dental or oral surgeries, it is essential to emphasize the potential for a difficult airway and ensure that the procedure is performed in a setting equipped to handle complications. If these avenues are not successful, the hospital operator can be asked to forward you to the "anesthesiologist on call," who may be able to assist in pointing you in the correct direction.

For these patients, generally, it is strongly recommended that all but the most minor procedures performed under local anesthesia be performed in a hospital setting—ideally at an institution with experience caring for patients with SD. Even experienced practitioners can experience difficulties in an out-of-hospital setting due to the lack of advanced equipment and backup support from knowledgeable individuals in an emergency. In an office setting, “backup” often is an ambulance crew.

In summary, when facing surgery, the perioperative period can be a daunting and stressful time for patients with SD and their families. However, it would be in the patient and family’s best interest to take charge of care, especially when finding the right anesthesiologist. Every member of the care team—the surgeons, medical specialists, or the anesthesiologists—must be on the same page. This requires proactive communication and meticulous planning. Patients and families should not underestimate the value of this preparation; it is essential for ensuring the safety and success of the procedure.

Identifying and Caring for the Medically Complex Patient: A Multidisciplinary Approach

4

Colleen P. Ditro

4.1 Introduction

Patients with a skeletal dysplasia (SD) often fall into the category of patients with medical complexities. The most common SDs managed at our facility are Morquio syndrome, achondroplasia, metatropic dysplasia, campomelic dysplasia, and diastrophic dysplasia. These patients may present with a multitude of medical issues or none, depending on the severity of the dysplasia. It is important to know what comorbidities typically accompany the dysplasia.

Patients with medical complexities often have chronic conditions that involve multiple organ systems and require a range of specialties and support. Generally, there are four features that characterize these patients: (1) chronic and severe health conditions, (2) substantial health service needs, (3) major functional limitations, and (4) high health care resource use [1, 2].

4.2 Structure and Function

Patients with medical complexities often undergo multiple elective surgeries related to their underlying diagnosis to improve function and enhance daily living activities. Preparing for surgery can be challenging due to their multiple health issues, which can impact their perioperative care, including the anesthetic management and postoperative recovery. Involvement of primary care specialties, such as general pediatrics, internal medicine, or family medicine, in the pre-operative care for patients with SD undergoing surgeries like spinal fusion for scoliosis, has been associated with better recommendations for the management of co-existing

C. P. Ditro (✉)
Clinical Operations, Orthopedics, Nemours Children's Hospital, Wilmington, Delaware, USA
e-mail: cditro@nemours.org

M. C. Theroux (ed.), *Anesthesia Care of Patients with Skeletal Dysplasia*,
https://doi.org/10.1007/978-3-031-92501-6_4

conditions [3]. This involvement has also been associated with shorter length of stay and decreased cost for the inpatient care [1, 3].

Parents with such complexity consider coordination to be the "critical unmet need related to accessing clinical care" [4]. A team led by an advanced practice registered nurse (APRN) can provide intensive medical care coordination for these patients and extensive support to their families. The APRNs familiarity with the hospital system and their relationship with the subspeciality groups enable them to provide safe and efficient patient care. Cady et al. [5] described the value of the APRN's enhanced scope of knowledge and practice for relationship-based care coordination. The APRN is also highly skilled yet less likely to be constrained by a limited environment because they are accustomed to providing direct care and coordination in a clinical setting.

Spine fusion with or without decompression is a major surgery often necessary for the majority of skeletal dysplasia patients. Berry et al. [6] reviewed 79 children aged 5–21 years undergoing spinal fusion and found that preanesthetic evaluation and clearance are labor-intensive processes often requiring last-minute care planning to optimize the child's perioperative safety. They found that all children required at least one last-minute care coordination activity and half of children required three or more activities. Most of the last-minute consults were related to co-existing health conditions. These investigators also found that a multidisciplinary approach early in the process can better identify and address the co-existing health conditions.

4.3 Multidisciplinary Care Coordination

Given the complexity of the different aspects of care, patients with skeletal dysplasia are likely to receive fragmented care across multiple facilities. Patient-centered care provided at one facility is ideal and leads to greater satisfaction of both the patients and providers. However, coordinating care for these very complex patients can be difficult to achieve. Challenges include care across multiple states, offices, and health care systems. Care coordination is an essential element of a transformed American health care delivery system that emphasizes optimal quality and cost outcomes, addresses family-centered care, and calls for partnership across providers, settings, and communities [7].

Managing an international and national SD program can be multi-faceted. Patients arrive from different parts of the country, and some from abroad, often already receiving fragmented care. When patients arrive either for a clinic visit or for surgery, it is imperative that all the consults and imaging studies are completed. When preparing for surgery, parents need to feel comfortable and confident that their child has been well prepared in both medical and psychological ways.

Our skeletal dysplasia team evaluated patients with SD from multiple states in the United States. In 2023, a total of 1347 patients were evaluated in clinic: 216 new patients and 1131 follow-up patients. The chart below separates the patients by states (Fig. 4.1). Roughly 200–250 patients undergo surgery and anesthesia, mostly

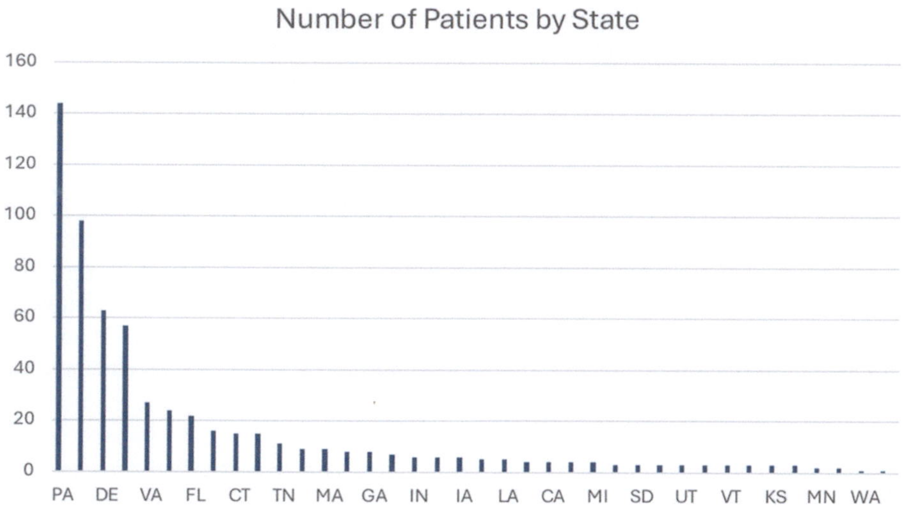

Fig. 4.1 Number of patients by state

for orthopedic and neurosurgical procedures. These data include patients who require anesthesia or deep sedation for medical imaging procedures as well.

For example, an 8-year-old female with Morquio syndrome traveling several states to be evaluated in the skeletal dysplasia clinic who is scheduled to have hip surgery 2 days after the evaluation needs to be thoroughly reviewed ahead of her arrival to the facility. A typical patient diagnosed with Morquio A syndrome may have cardiac and pulmonary issues. In addition, patients with Morquio A must be screened for cervical instability and/or cervical stenosis. If these evaluations are not completed on time prior to surgery, they carry the risk of having possible devastating complications, including spinal cord injury.

At our facility, multiple patients are reviewed in a weekly Skeletal Dysplasia Team meeting for various reasons. This meeting includes several specialties, such as genetics, orthopedics, neurosurgery, anesthesia, pre-operative nurse practitioners, physical therapy, and, if necessary, pulmonology, otolaryngology, urology, and cardiac. At this meeting, imaging studies are reviewed for both new patients scheduled for clinic visits and for patients seeking follow-up of their previous visits This meeting, attended by anesthesiologists, provides a much-needed venue to discuss concerns and aspects of care that need greater consideration and coordination. For example, a patient with Morquio A may need further airway evaluation, which may include having an otolaryngologist available on the day of surgery. Or, the procedure and the patient status are such that neuromonitoring may be necessary and must be arranged in advance. These are but a few examples of the multitude of necessary care components.

In addition, there is a monthly, multidisciplinary skeletal dysplasia coordination rounds that includes nursing, physical therapy, occupational therapy, social work, child life, case management, respiratory therapy, and rehabilitative therapy. This meeting discusses upcoming surgeries and the patients' needs for successful recovery, discharge, and a possible long day of travel home if the patient is not local to the hospital. Some patients undergo extensive surgery that requires care coordination for equipment needs such as wheelchairs and walkers. Patients undergoing hip reconstruction will need to stay in the rehabilitative department, which requires careful evaluation of insurance and staffing needs. Another meeting involves a discussion with the surgeon, the advanced practice provider, and administration to review the upcoming surgical cases for the appropriate coding and equipment.

All three of these multidisciplinary meetings are highly valued, where multiple teams work together to improve the pre-operative care coordination between patients, providers, and ancillary support personnel. Each team member is an integral part of the health care team and works to improve patient outcomes and increase access to care.

4.4 Summary

Patients with SD can be medically and surgically complex, especially when preparing for surgery. There are several opportunities to optimize the care of patients, including reviewing images and discussing the patient's known comorbidities prior to their arrival in clinic for an evaluation. This process should continue into the pre-operative stage with multiple consults, if needed. A review of the patient's surgical procedure before surgery with the surgeon, and a review of the patient's post-surgical needs are essential elements of the coordination meeting. All of this requires a multidisciplinary team approach to provide the highest level of care and ensure good surgical outcomes.

References

1. Berry JG, Glotzbecker M, Rodean J, Leahy I, Cox J, Singer SJ, et al. Perioperative spending on spinal fusion for scoliosis for children with medical complexity. Pediatrics. 2017;140(4):e20171233.
2. Agrawal R, Hall M, Cohen E, Goodman DM, Kuo DZ, Neff JM, et al. Trends in health care spending for children in medicaid with high resource use. Pediatrics. 2016;138:e20160682.
3. Rappaport DI, Adelizzi-Delany J, Rogers KJ, Jones CE, Petrini ME, Chaplinski K, et al. Outcomes and costs associated with hospitalist comanagement of medically complex children undergoing spinal fusion surgery. Hosp Pediatr. 2013;3:233–41.
4. Allshouse C, Comeau M, Rodgers R, Wells N. Families of children with medical complexity: a view from the front lines. Pediatrics. 2018;141:S195–201.
5. Cady RG, Kelly AM, Finkelstein SM, Looman WS, Garwick AW. Attributes of advanced practice registered nurse care coordination for children with medical complexity. J Pediatr Health Care. 2014;28:305–12.
6. Berry JG, Glaspy T, Eagan B, Singer S, Glader L, Emara N, et al. Pediatric complex care and surgery comanagement: preparation for spinal fusion. J Child Health Care. 2020;24:402–10.
7. Council on Children with Disabilities and Medical Home Implementation Project Advisory Committee. Patient- and family-centered care coordination: a framework for integrating care for children and youth across multiple systems. Pediatrics. 2014;133:e1451–60.

Part II

Individual Skeletal Dysplasias

Achondroplasia

5

Abraham Oommen

5.1 Genetics

Achondroplasia (ACH) is an autosomal dominant mutation of the *FGFR3* gene. This mutation is fully penetrant and shows only marginal variability of expression [1]. Roughly 250,000 people worldwide are affected by ACH [2], which is about 0.003–0.004% of the global population [3]. Novel mutations of the genetic code account for 80% of cases of ACH, and the remaining 20% of cases occur because of the autosomal dominant inheritance pattern [4]. This means that the overwhelming majority of those with ACH are born to parents of average stature. An average stature individual whose partner has ACH has a 50% chance of having offspring with average stature and a 50% chance of having affected offspring. The children of two individuals with ACH will have a 50% chance of having ACH and a 25% chance of having average stature. The remaining 25% of the time, the offspring will be homozygous for ACH. Homozygosity for ACH is almost always lethal [5] and has a remarkably similar phenotype to thanatophoric dysplasia (TD). A recently published case report by one of our colleagues challenges the lethality of TD by describing the case of a boy with TD who is currently 11 years old [6].

ACH can be complicated by other common genetic disorders, one of which is Down syndrome (DS). DS occurs more commonly with advanced maternal age whereas ACH has a higher incidence with advanced paternal age [1]. Seven instances of this co-occurrence have been reported in the literature [7]. These genetic conditions have comorbidities that seemingly overlap, resulting in additive pathology that can be difficult to manage perioperatively. Comorbid

A. Oommen (✉)
Department of Anesthesiology and Perioperative Medicine, Nemours Children's Hospital, Wilmington, Delaware, USA

Anesthesiology and Pediatrics, Sidney Kimmel Medical College, Thomas Jefferson University, Philadelphia, Pennsylvania, USA
e-mail: Abraham.Oommen@nemours.org

M. C. Theroux (ed.), *Anesthesia Care of Patients with Skeletal Dysplasia*,
https://doi.org/10.1007/978-3-031-92501-6_5

conditions include obstructive sleep apnea (OSA), cranio-cervical spine pathology, cervical spinal cord instability, restrictive pulmonary disease, difficult intravenous (IV) access, and challenging airways for intubation. All of these comorbid conditions complicate perioperative management. It is imperative that appropriate experience and skill in management of patients with ACH be demonstrated prior to caring for a patient with a double disorder.

It has been many centuries since ACH was first described, but only recently has the molecular pathogenesis been discovered. In 1994, Shiang and colleagues described the *FGFR3* gene mutation that causes ACH [8]. Almost all mutations in the *FGFR3* gene arise in the same nucleotide pair and result in a glycine to arginine substitution (G380R) in the FGFR3 protein [9]. Newer evidence is pointing to advanced paternal age as the primary culprit for the de novo genetic mutation. Unaffected fathers in their 50s are ten times more likely to have offspring with a de novo *FGFR3* mutation when compared with fathers in their 20s [10]. ACH is one of a small number disorders characterized by recurrence, autosomal dominance, male bias, and paternal aging (RAMP disorders [1]).

5.2 Phenotype

The primary phenotypical observation in ACH is short stature, which may not be readily apparent at birth. The average adult height is about 4 feet for both men and women with ACH. The short stature in ACH results from relative shortening of the proximal limbs, a phenotype referred to as rhizomelia [11]. The humerus and the femur are disproportionally affected leading to proximal limb shortening (Figs. 5.1 and 5.2).

Fig. 5.1 Typical phenotype of an infant with ACH showing frontal bossing, macrocephaly, rhizomelic disproportion, and flat nasal bridge. Mild midface hypoplasia can also be seen. (Consent for image obtained)

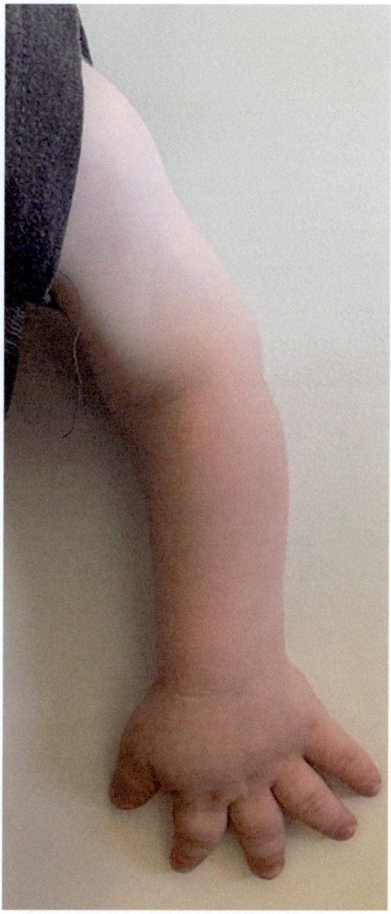

Fig. 5.2 Rhizomelia of upper extremity

Because of rhizomelic disproportion, there are commonly skin folds and creases in the proximal arms and legs [1]. Macrocephaly is common with frontal bossing, and the midface is hypoplastic resulting from cartilage growth defects at the skull base (Fig. 5.1). There are varying degrees of midface hypoplasia with some much less severe than others. Hypoplasia of bony cartilage in the face results in facial flattening and a flat nasal bridge [1]. By infancy, almost all individuals with ACH develop thoracolumbar kyphosis, which is not present at birth [12].

Other features include short hands and fingers, known as brachydactyly. Trident hand (Fig. 5.3) is a term that has been used to describe the phenotypical appearance

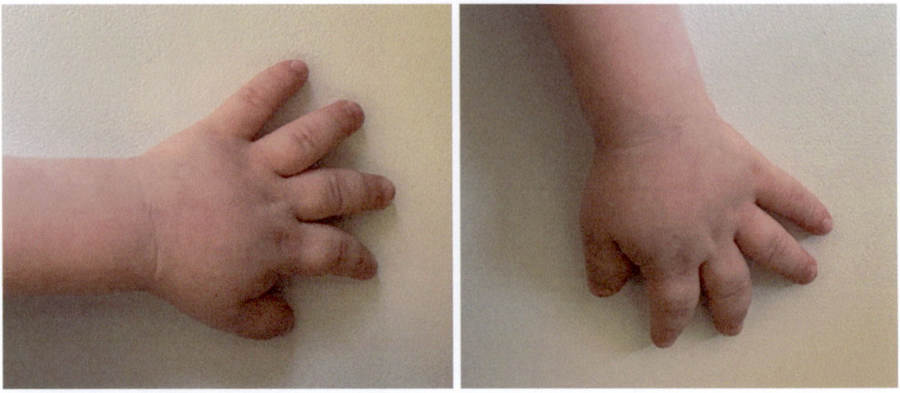

Fig. 5.3 Classic trident hand

of the distal upper extremity in ACH. Bowing of extremities resulting in varus deformities is common. Hypotonia is common in infancy as well and often results in delayed meeting of milestones.

5.3 Radiological Findings

It is challenging to make a radiological in utero diagnosis of ACH until about the third trimester using ultrasound. Key features are femur length below the fifth percentile, trident hand, large forehead, and a flat nasal bridge. A combination of these findings can lead to definitive diagnosis. Diagnostically helpful features include the following: short, robust tubular bones; squared-off iliac wings; flat, horizontal acetabula; marked narrowing of the sacrosciatic notch; a characteristic proximal femoral radiolucency; narrowing of the interpediculate distance of the caudal spine; and short proximal and middle phalanges [13]. The following group of images and descriptions are classic radiological findings in individuals with ACH (Figs. 5.4, 5.5, 5.6, and 5.7).

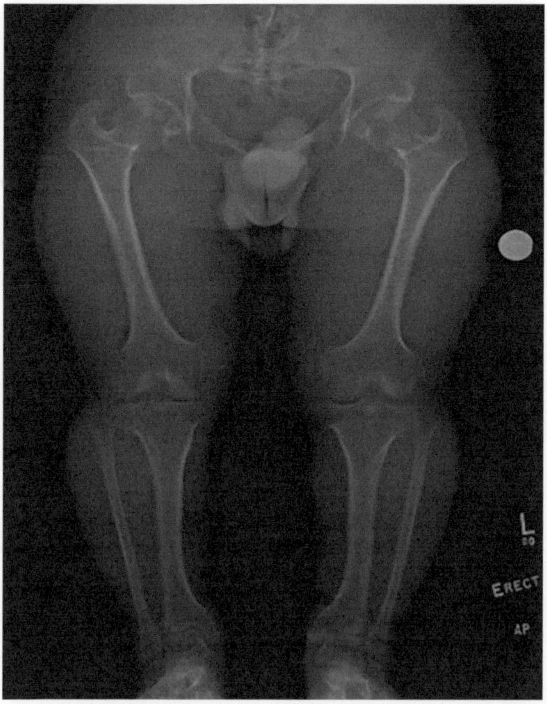

Fig. 5.4 Radiographic evidence of shortened long bones in an adolescent, most prominently the femur. *FGFR3* affects long bone growth through a negative feedback loop that controls the differentiation of chondrocytes in the growth plate [14, 15]. Premature chondrocyte differentiation causes the proliferation stage of the chondrocytes to be shortened, which results in a decrease in the overall length of the long bones [15]. Note the long fibula in comparison with the tibia

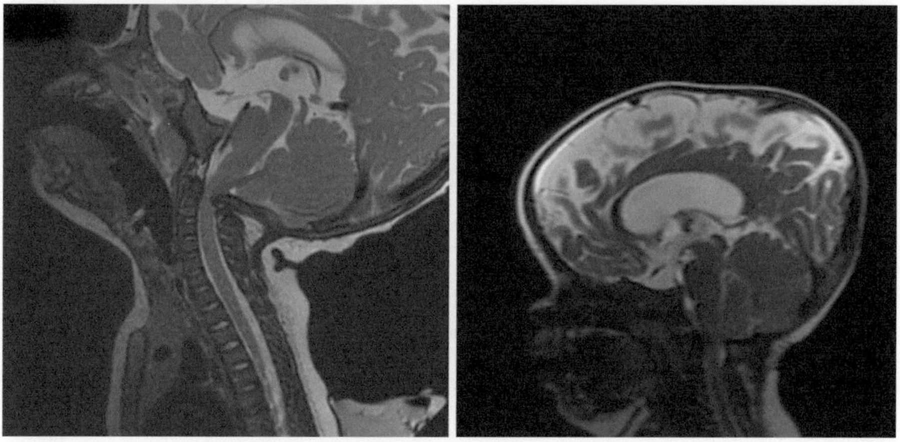

Fig. 5.5 Severe narrowing of the cervical canal at the level of the foramen magnum, which worsens with extension. Central apnea, hypotonia, hyperreflexia, weakness, and developmental delays are all associated with foramen magnum stenosis (FMS). Image on the right is status post decompression of the foramen magnum (further described later in this chapter)

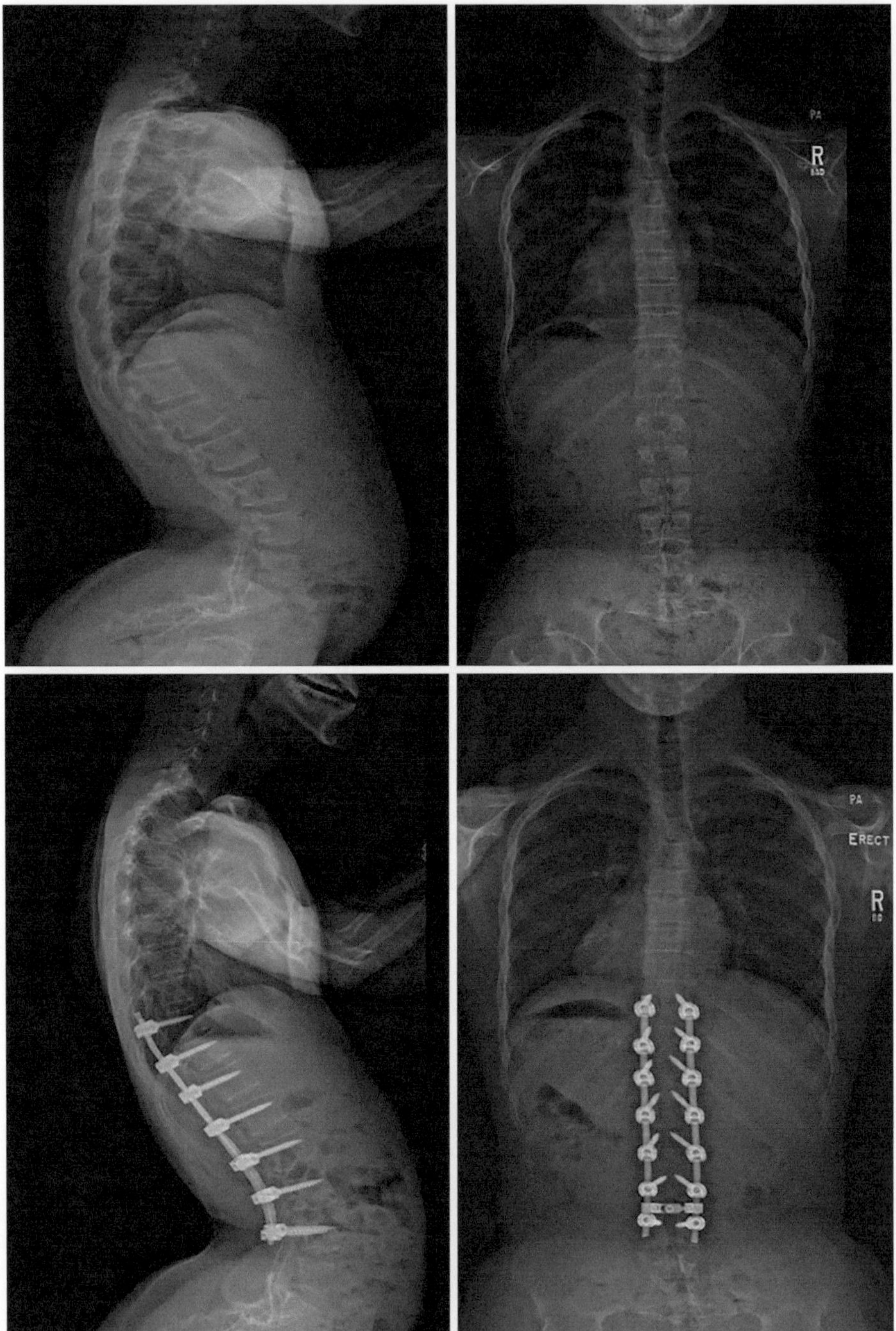

Fig. 5.6 Before and after radiographs of a teenager with ACH who had a T10-L4 posterior spinal fusion and an L1–L4 laminectomy for symptomatic spinal stenosis. Symptoms included lower extremity paresthesias and rapid muscular fatigue after short distances. Note posterior vertebral body scalloping on the lateral preoperative film, which is an exaggerated concavity of the veterbral body

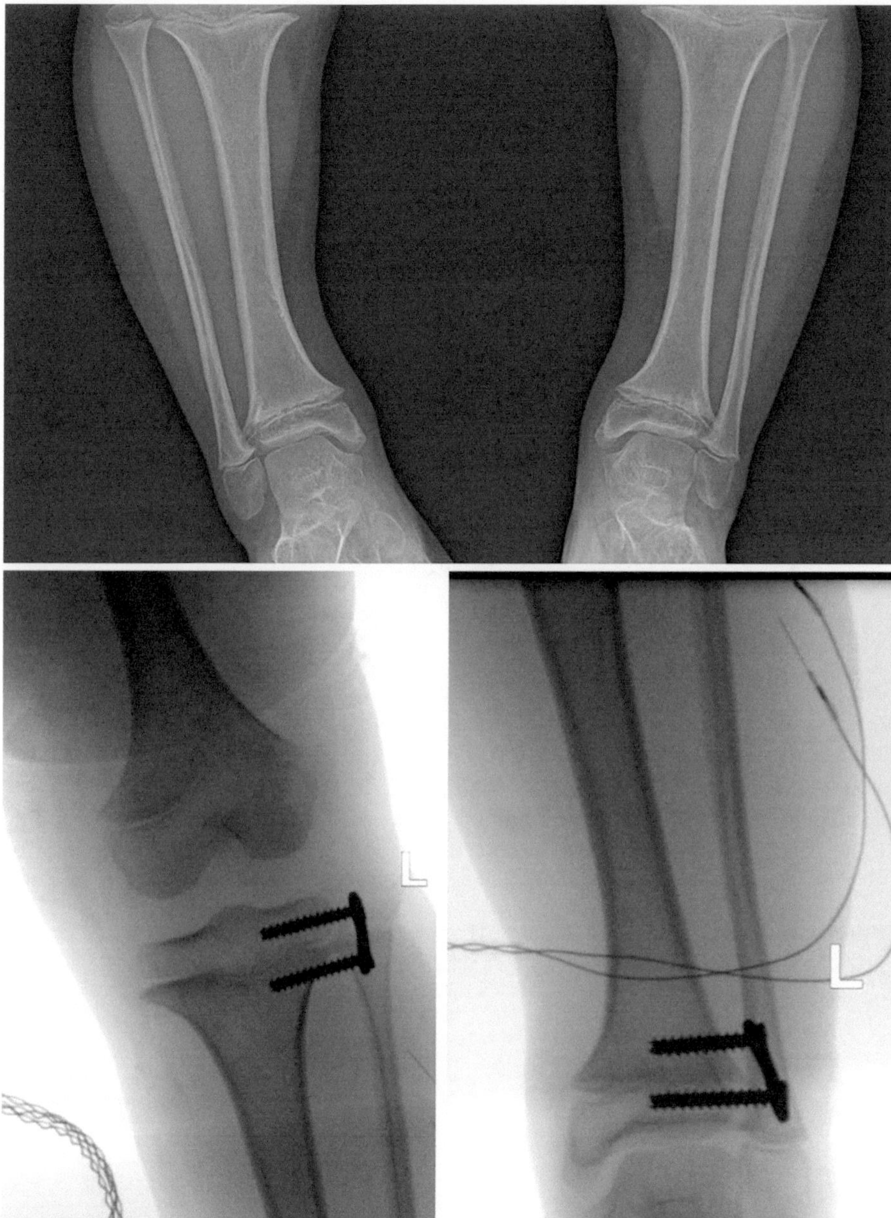

Fig. 5.7 The top radiograph shows an adolescent with bilateral genu varum, which is bowing of the legs outward at the knees. The bottom radiographs show the placement of knee (bottom left) and ankle (bottom right) hemiepiphysiodesis growth modulation plates with cannulated screws. The goal of these growth modulation plates is to correct the genu varum. Intraoperative neuromonitoring leads and wires are also visible in the bottom radiographs. In patients without ACH, these growth plates routinely are used on a single extremity for limb length discrepancy correction

5.4 Comorbidities

There are numerous comorbidities associated with ACH that are important when considering anesthetic care and management of these patients. Here, we will take a systems-based approach to describe many of the comorbid conditions seen in ACH. We will then group comorbidities by the age of presentation (Table 5.1). A good understanding of these issues and the timeline of their presentation is helpful when designing a trouble-free anesthetic.

5.4.1 General Appearance

Most individuals with ACH have large heads and frontal bossing (see phenotypical description earlier in this chapter for more details). Obesity with aging is a significant concern and the risk of obesity is higher in the ACH population. Obesity makes vascular access, airway management, and perioperative care much more challenging.

5.4.2 Central Nervous System

Foramen magnum stenosis (FMS) in infancy leads to cervicomedullary cord compression (CMCC), which causes central apnea and can result in sudden infant death. The risk of death associated with FMS, on the conservative side, is 2–3% [1], but some quote a higher risk of up to 7.5% [16, 17]. The association of sudden death in infants with ACH attributable to sequalae from FMS-related CMCC first was noted in the 1980s [18]. Numerous infants died aged 3–11 months, leading to this

Table 5.1 Summary of comorbidities in ACH grouped by age

Infants & toddlers	**Developmental delays**
	Hypotonia
	Sleep-disordered breathing
	Foramen magnum stenosis
	Thoracolumbar kyphosis
	Cervicomedullary cord compression
Childhood	Otitis media
	Conductive hearing loss
	Genu varum
	Obstructive sleep apnea
	Poor maxillary arch growth
	Dental malocclusion
Adolescence & adulthood	Psychosocial acceptance
	Spinal stenosis
	Obesity
	Hypertension
	Chronic pain
	Pregnancy-related care

conclusion [18]. Likely, a combination of factors played a role in these deaths, including combined central and obstructive sleep apnea. Early diagnosis and treatment in the first 2 years of life is beneficial [19]. Diagnosis of symptomatic FMS can be challenging. Physical exam findings of lower extremity clonus and hyperreflexia are sometimes noted. Polysomnography to define central apnea is useful in centers that are skilled at performing the test on infants. Cranio-cervical magnetic resonance imaging (MRI) is the gold standard for assessment of CMCC. A review of the literature shows that one of every ten infants with ACH will require surgical decompression for FMS [1].

Individuals with ACH have normal intelligence. There is an increased risk of hydrocephalus in the first 2 years of life, requiring shunt placement. Head circumference should be monitored monthly during the first year of life and compared with norms for infants with ACH [20]. The incidence of shunt placement has been reported as 4.3% [21]. The etiology of hydrocephalus is somewhat unclear but can be attributed to either FMS blocking outflow or intracranial venous hypertension resulting in decreased cerebrospinal fluid resorption [22–24].

Mental health and social acceptance are important issues when caring for individuals with ACH, especially as they age. Acceptance by peers and age-appropriate treatment during social interactions is a significant issue that children and teens with ACH must face.

5.4.3 Ear, Nose, and Throat

The most common otolaryngologic problem associated with ACH is recurrent otitis media, which must be treated to prevent hearing loss [11]. Hearing loss can lead to subsequent speech and learning disabilities. Overgrowth of adenoids and anatomical differences in the ear canal cause recurrent bouts of otitis media. Surgical placement of myringotomy tubes is the best treatment option.

Midface hypoplasia, a narrow upper airway, and crowding of the oropharynx often lead to OSA. If OSA is not present, it is likely that there is some degree of sleep-disordered breathing (SDB). The continuum of SDB ranges from mild snoring to upper airway resistance syndrome (UARS) to OSA. The primary difference between UARS and OSA is the lack of true apnea in UARS. Left untreated, UARS almost always progresses to OSA. Obstructive sleep apnea is especially worrisome when combined with central apnea caused by CMCC. The risk of OSA at any age is between 30% and 40% [25–27]. In toddlers and school-aged children, hypertrophy of lymphatic tissues worsens oropharyngeal crowding, further narrowing the upper airway. In adolescents and adults, development of obesity is associated with worsening SDB and progression to OSA [1]. The primary treatment for UARS and OSA is adenotonsillectomy. Other treatment options include continuous positive airway pressure (CPAP), although compliance is an issue, and uvulopalatopharyngoplasty (UPPP), which aims to rearrange oropharyngeal tissues to increase airway size and reduce tissue collapse. In rare circumstances, a tracheostomy is needed to bypass the upper airway; however, we rarely see this at our institution.

Orthodontic issues related to midface hypoplasia are common. Issues with the palate and bite dynamics are frequent. Assessment by an orthodontic specialist is important for children with ACH because of poor maxillary arch growth [28]. If after pharyngeal surgery there is still residual OSA in individuals with ACH, maxillary orthodontic expansion and maxillary non-surgical protraction can be considered to increase upper airway volume [29, 30].

5.4.4 Cardiopulmonary

Small chest size and an overly compliant rib cage can cause respiratory issues in infants. A pediatric pulmonologist should follow infants with ACH.

The heart is unaffected from a physiologic standpoint. Obstructive sleep apnea that is poorly controlled for many years can lead to cardiac disease from pulmonary hypertension. Obstructive sleep apnea also is associated with obesity, heart disease, and diabetes. There is evidence of greater risk of death secondary to cardiovascular conditions. The mortality rate secondary to heart disease is twice as high for those with ACH compared with average stature individuals [17]. Hypertension is also prevalent in older patients with ACH.

5.4.5 Gastrointestinal

Reflux in infancy combined with oro-motor hypotonia can result in slow weight gain and delayed milestones; however, this problem is usually self-limited. It is rare to need an assistive feeding device such as a gastric tube. In adolescents and adults, obesity is a prevalent issue, and there have been successful reports of bariatric surgery in individuals with ACH. At the time of writing, GLP-1 agonists are starting to become popular for weight loss. Time will tell how this can be utilized in those with ACH.

5.4.6 Musculoskeletal

Rhizomelia of the long bones is a classic feature of ACH. Genu varum, or outward bowing of the legs, occurs in 40–70% of patients with ACH [31]. Common indications for surgical intervention with growth modulation plates include pain, gait abnormalities, and joint instability. See the radiological findings section of this chapter for a more detailed description of these musculoskeletal findings. Most infants with ACH are hypotonic [32]. Joint hypermobility is common in ACH with one notable exception: the elbow. There is limited elbow extension, which further shortens the reach of an individual with ACH. All individuals with ACH have substantially diminished caliber of the spinal canal along its entire length [33, 34]. Thoracolumbar kyphosis is present in early life but improves with age. Lumbar lordosis worsens with age and often results in neurogenic claudication requiring laminectomy and posterior spinal fusion.

Limb-lengthening surgery is a controversial topic among both patients with ACH and orthopedic surgeons who specialize in skeletal dysplasias. The complexity and recovery time must be weighed against functional improvements and the patient's own cosmetic desires. Psychosocial acceptance and well-being undoubtedly play a role in the decision-making process. Acanthosis nigricans and craniosynostosis are associated with mutations of FGFR3 [35, 36]. When caring for patients with ACH, these pathologies may be encountered although rare.

5.4.7 Reproductive

Cesarean section (CS) is recommended for all pregnant women with ACH when delivery is after 32 weeks because of the shape and size of the pelvis [37, 38]. General or neuraxial anesthetics can be used for delivery depending on the clinical course and experience of the anesthesiologist. Delivery should take place at a well-equipped hospital with an intensive care unit, a blood bank, and personnel comfortable with the management of patients with ACH [37, 38].

5.5 Anesthetic Care

The anesthetic care of a patient with ACH is important and can be broken down into three phases. The components of perioperative care include pre-, intra-, and postoperative care. All phases of care are equally important, and we will discuss each in detail. The goal of this portion of this chapter is to describe our institution's perioperative management of patients with ACH. Table 5.2 offers a summary of the most common surgical procedures, and Table 5.3 offers a standard anesthetic plan that will work for the majority of cases.

Table 5.2 Common procedures and imaging studies in ACH

Surgical procedures	Imaging/miscellaneous procedures
Myringotomy tubes	MRI brain
Adenotonsillectomy	MRI cervical spine: Flexion/extension
Foramen magnum decompression	MRI thoracolumbar spine
Ventriculoperitoneal shunt	CT scan
Third ventriculostomy	Dental/orthodontic procedures
Growth modulation plates (eight plates)	
Laminectomy	
Posterior spinal fusion	
Cesarean section	

Table 5.3 Anesthetic plan for common procedures in ACH

Procedure	Anesthetic plan
MRI scans[a]	Deep sedation (natural airway)
Myringotomy tubes MRI scans Growth modulation plates (Hemiepiphysiodesis)	GA, LMA, IV ×1
Adenotonsillectomy Ventriculoperitoneal shunt Third ventriculostomy Dental & orthodontics	GA, ETT, IV ×1
Adenotonsillectomy[b] Growth modulation plates (Hemiepiphysiodesis)[b]	TIVA, ETT, neuromonitoring, IV ×2
Foramen magnum decompression Posterior spinal fusion Laminectomy	TIVA, ETT, neuromonitoring, IV ×2, arterial line
Cesarean section	GA vs. neuraxial techniques

[a] Possible for MRI scans without evidence of SDB except MRI cervical spine with flexion and extension

[b] Adenotonsillectomy and hemiepiphysiodesis are listed twice because depending on the extent of patient disease they could require additional monitoring and a more involved anesthetic

5.5.1 Preoperative

Patients come to our institution from across the country, and we prepare for them accordingly. A weekly multidisciplinary meeting is held to discuss upcoming cases, preoperative testing needed, and potential perioperative challenges. We have licensed health care workers who are also members the skeletal dysplasia community play an active role in this multidisciplinary team.

A thorough preoperative evaluation is important. Any existing anesthetic records with airway documentation should be reviewed. Important information gathered includes ease of ventilation, seating/size of laryngeal mask airway (LMA), and devices used for successful intubation. These patients are frequent fliers and should have anesthetic records available for review.

Patients with ACH travel great distances from across various parts of the country for expert care; however, this fact should not supersede patient safety. If a patient is under the weather, their surgery should be rescheduled like any other patient. These patients have less pulmonary reserve than the general population, so caution must be exercised when treating a patient with a recovering or acute upper respiratory infection.

Healthy patients benefit from following established enhanced recovery after surgery (ERAS) protocols. The two most common applications of this would be for adenotonsillectomy and posterior spinal fusion. For example, our adolescent patients with idiopathic scoliosis receive a dose of gabapentin, acetaminophen, and celecoxib in the preoperative holding area upon being checked in for surgery. We have started to apply a similar ERAS regimen to our patients with ACH for spinal fusion. An ERAS protocol often implies no individualized modifications to the regimen, which is not advisable in most patients with skeletal dysplasia including ACH.

5.5.2 Intraoperative

The first decision that needs to be made is how to induce general anesthesia (GA). Distinct craniofacial features can lead to difficulty with mask ventilation and airway management. As pediatric anesthesiologists, we are generally comfortable with the inhalation induction of GA prior to IV line placement. This modality of inducing GA is a skill that is honed over many years of experience and should be not attempted in patients with ACH without proficiency. There will be a fast uptake of inhalational anesthetics because of decreased functional residual volume resulting in a shorter time to loss of consciousness in these children. Even when skilled in inhalation induction techniques, there are certain patients for whom we elect to start a pre-induction IV line. This is practitioner dependent, but teenagers and patients who are obese benefit from awake or moderately sedated IV placement to enhance safety. Elsewhere, IV lines are placed awake in all patients such as pediatric emergency departments, pediatric wards, and adult facilities. Should there be any concern for difficulty with ventilation or intubation, it is highly advisable to secure IV access prior to the induction of GA. Loss of airway compounded by a lack of IV access is a situation that every anesthesiologist should avoid.

After induction of GA, an appropriately sized LMA or endotracheal tube (ETT) is placed. Direct or indirect laryngoscopy can be used to secure the airway when placing an ETT. We often opt to proceed directly to video laryngoscopy with our first attempt to maximize first-pass success rate. Magill forceps are often needed to pull the tongue forward to create room for the video laryngoscope [39]. A video laryngoscope with a hyperangulated blade and flexible fiberoptic bronchoscope (FOB) are the most likely to succeed consistently and reliably when attempting to intubate the trachea in challenging scenarios [40, 41]. LMA use as a conduit to perform FOB intubations is another technique that allows for intubation while maintaining ventilation during the intubation process [42, 43].

Occasionally, a natural airway is used for deep sedation. This is rarely done in the operating room but can be used as a technique for acquiring MRI images. The use of deep sedation with a natural airway is cautioned if there is evidence of SDB, UARS, or OSA.

Intravenous access can be a challenge in this patient population because of increased subcutaneous fat and redundant soft tissue. It is important to develop ultrasound-based vascular access skills to adequately care for these patients. At our institution, arterial lines are placed under ultrasound guidance increasing first-attempt success rate and minimizing time to anesthesia readiness.

Routine standard monitoring is appropriate for all patients. Because of rhizomelia of the upper extremities, finding an appropriate blood pressure cuff fit can be challenging. We often use the forearm for blood pressure readings if the upper arm is significantly disproportionate. Refer to the chapter on neurophysiologic monitoring for an in-depth summary of how, when, and why we choose to use this advanced monitoring technique. Our neurophysiology colleagues have extensive experience with patients with ACH and are extremely valuable members of the perioperative team.

Maintenance of anesthesia depends on the use of neuromonitoring. Total intravenous anesthesia (TIVA) is used when the case is being neuromonitored. Most commonly, we use propofol as our sedative and remifentanil as our analgesic. We embrace the use of remifentanil because of its short context-sensitive half time, which allows patients to be extubated without the respiratory depressant effects of long-acting narcotics. Detailed information related to the intraoperative care including safe extubation in the operating room is described in the chapter titled "Anesthesia Care of Spine Fusion." A sevoflurane-based anesthetic is utilized during routine cases without neuromonitoring.

Our goal is to extubate all patients in the operating room. If there is a substantial risk of reintubation because of edema from resuscitation, patients are taken to the intensive care unit intubated with the goal of rapid extubation. Numerous factors facilitate extubation. Regional analgesic techniques allow us to minimize narcotic use. Judicious crystalloid usage prevents global edema. Anti-inflammatory steroid use such as dexamethasone and methylprednisolone helps to minimize airway edema and reactivity. All the above aid in timely extubation of our patients.

Special considerations for obstetrical surgery are warranted. Deliveries in women with ACH must be via CS because of uniform narrowing of the pelvis and cephalopelvic disproportion [44]. Neuraxial techniques have been reported in patients with ACH [45]. Unusual difficulties have been described [46–48] with neuraxial techniques, which is not surprising given the anatomical spinal canal differences compared to those of average stature. Anesthetic management for CS remains a controversial topic, and no standard of care has been reported [49]. There have been anecdotal reports of underdosed spinal medications in patients for CS resulting in traumatic experiences for the mothers. Low dose intrathecal local anesthetic is likely an attempt to avoid a high spinal and the risks associated with GA in a pregnant mother with ACH. Although unlikely, the anesthesiologist should be prepared for GA and supportive measures in the instance of a high spinal. Spinal medications should be dosed appropriately for adequate coverage as the primary anesthetic. A spinal catheter is another anesthetic option where the local anesthetic is administered intrathecally and can be titrated to affect.

5.5.3 Postoperative Care

Pain management and control of nausea/emesis are the most important factors in the immediate postoperative period. Our acute pain management service (APMS) follows patients who are admitted overnight. The APMS uses a multimodal approach to control pain. Refer to the chapter on pain management and regional anesthesia for further details. See Table 5.4 for a summary of our

Table 5.4 Multimodal pain plan

APMS dosing regimen	
Acetaminophen (15 mg/kg) Max 1000 mg	Scheduled every 6 h for 24 h followed by as-needed dosing
Ketorolac (0.5 mg/kg) Max 15 mg	Scheduled every 6 h for 48 h followed by as-needed ibuprofen/naproxen
Oxycodone (0.1 mg/kg) Max 5 mg	Every 6 h as needed for moderate to severe pain
Diazepam (0.1 mg/kg) Max 2.5 mg	Every 6 h as needed for muscle spasms
Ropivacaine infusion (0.1–0.4 mg/kg)	Via nerve block catheter if applicable
Gabapentin Clonidine Ketamine	At the discretion of the APMS, use based on extent of surgery and patient comorbidities

multimodal analgesic regimen. Adequate pain control leads to decreased hospital length of stay, improves patient satisfaction, and allows for quicker participation in physical therapy, thereby hastening recovery. Having an APMS team manage these patients is of utmost importance to minimize postoperative respiratory events, limit the development of opioid dependence, and avoid addiction to controlled substances.

References

1. Pauli RM. Achondroplasia: a comprehensive clinical review. Orphanet J Rare Dis. 2019;14:1.
2. Ireland PJ, Pacey V, Zankl A, Edwards P, Johnston LM, Savarirayan R. Optimal management of complications associated with achondroplasia. Appl Clin Genet. 2014;7:117–25.
3. Moffitt KB, Abiri OO, Scheuerle AE, Langlois PH. Descriptive epidemiology of selected heritable birth defects in Texas. Birth Defects Res A Clin Mol Teratol. 2011;91:990–4.
4. Orioli IM, Castilla EE, Scarano G, Mastroiacovo P. Effect of paternal age in achondroplasia, thanatophoric dysplasia, and osteogenesis imperfecta. Am J Med Genet. 1995;59:209–17.
5. Pauli RM, Conroy MM, Langer LO Jr, McLone DG, Naidich T, Franciosi R, et al. Homozygous achondroplasia with survival beyond infancy. Am J Med Genet. 1983;16:459–73.
6. Carroll RS, Duker AL, Schelhaas AJ, Little ME, Miller EG, Bober MB. Should we stop calling thanatophoric dysplasia a lethal condition? A case report of a long-term survivor. Palliat Med Rep. 2020;1:32–9.
7. de Azevedo Moreira LM, Matos MA, Schiper PP, Carvalho AF, Gomes IC, Rolemberg JC, et al. Co-occurrence of achondroplasia and down syndrome: genotype/phenotype association. Birth Defects Res A Clin Mol Teratol. 2010;88:228–31.
8. Shiang R, Thompson LM, Zhu YZ, Church DM, Fielder TJ, Bocian M, et al. Mutations in the transmembrane domain of FGFR3 cause the most common genetic form of dwarfism, achondroplasia. Cell. 1994;78:335–42.

9. Bellus GA, Hefferon TW, Ortiz de Luna RI, Hecht JT, Horton WA, Machado M, et al. Achondroplasia is defined by recurrent G380R mutations of FGFR3. Am J Hum Genet. 1995;56:368–73.
10. Daugherty A. Achondroplasia: Etiology, clinical presentation, and management. Neonatal Netw. 2017;36:337–42.
11. Ornitz DM, Legeai-Mallet L. Achondroplasia: development, pathogenesis, and therapy. Dev Dyn. 2017;246:291–309.
12. Pauli RM, Breed A, Horton VK, Glinski LP, Reiser CA. Prevention of fixed, angular kyphosis in achondroplasia. J Pediatr Orthop. 1997;17:726–33.
13. Langer LO Jr, Baumann PA, Gorlin RJ. Achondroplasia: clinical radiologic features with comment on genetic implications. Clin Pediatr (Phila). 1968;7:474–85.
14. Chitty LS, Griffin DR, Meaney C, Barrett A, Khalil A, Pajkrt E, et al. New aids for the non-invasive prenatal diagnosis of achondroplasia: dysmorphic features, charts of fetal size and molecular confirmation using cell-free fetal DNA in maternal plasma. Ultrasound Obstet Gynecol. 2011;37:283–9.
15. Placone J, Hristova K. Direct assessment of the effect of the Gly380Arg achondroplasia mutation on FGFR3 dimerization using quantitative imaging FRET. PLoS One. 2012;7:e46678.
16. Hecht JT, Francomano CA, Horton WA, Annegers JF. Mortality in achondroplasia. Am J Hum Genet. 1987;41:454–64.
17. Wynn J, King TM, Gambello MJ, Waller DK, Hecht JT. Mortality in achondroplasia study: a 42-year follow-up. Am J Med Genet A. 2007;143A:2502–11.
18. Pauli RM, Scott CI, Wassman ER Jr, Gilbert EF, Leavitt LA, Ver Hoeve J, et al. Apnea and sudden unexpected death in infants with achondroplasia. J Pediatr. 1984;104:342–8.
19. Cheung MS, Irving M, Cocca A, Santos R, Shaunak M, Dougherty H, et al. Achondroplasia foramen magnum score: screening infants for stenosis. Arch Dis Child. 2021;106:180–4.
20. Trotter TL, Hall JG. American Academy of Pediatrics Committee on genetics. Health supervision for children with achondroplasia. Pediatrics. 2005;116:771–83.
21. King JA, Vachhrajani S, Drake JM, Rutka JT. Neurosurgical implications of achondroplasia. J Neurosurg Pediatr. 2009;4:297–306.
22. Bruhl K, Stoeter P, Wietek B, Schwarz M, Humpl T, Schumacher R, et al. Cerebral spinal fluid flow, venous drainage and spinal cord compression in achondroplastic children: impact of magnetic resonance findings for decompressive surgery at the cranio-cervical junction. Eur J Pediatr. 2001;160:10–20.
23. Pierre-Kahn A, Hirsch JF, Renier D, Metzger J, Maroteaux P. Hydrocephalus and achondroplasia. A study of 25 observations. Childs Brain. 1980;7:205–19.
24. Steinbok P, Hall J, Flodmark O. Hydrocephalus in achondroplasia: the possible role of intracranial venous hypertension. J Neurosurg. 1989;71:42–8.
25. Sisk EA, Heatley DG, Borowski BJ, Leverson GE, Pauli RM. Obstructive sleep apnea in children with achondroplasia: surgical and anesthetic considerations. Otolaryngol Head Neck Surg. 1999;120:248–54.
26. Collins WO, Choi SS. Otolaryngologic manifestations of achondroplasia. Arch Otolaryngol Head Neck Surg. 2007;133:237–44.
27. Afsharpaiman S, Saburi A, Waters KA. Respiratory difficulties and breathing disorders in achondroplasia. Paediatr Respir Rev. 2013;14:250–5.
28. Stephen L, Holmes H, Roberts T, Fieggen K, Beighton P. Orthodontic management of achondroplasia in South Africa. S Afr Med J. 2005;95:588–9.
29. Zaffanello M, Cantalupo G, Piacentini G, Gasperi E, Nosetti L, Cavarzere P, et al. Sleep disordered breathing in children with achondroplasia. World J Pediatr. 2017;13:8–14.
30. Machado-Junior AJ, Zancanella E, Crespo AN. Rapid maxillary expansion and obstructive sleep apnea: a review and meta-analysis. Med Oral Patol Oral Cir Bucal. 2016;21:e465–9.
31. Lee ST, Song HR, Mahajan R, Makwana V, Suh SW, Lee SH. Development of genu varum in achondroplasia: relation to fibular overgrowth. J Bone Joint Surg Br. 2007;89:57–61.
32. Reynolds KK, Modaff P, Pauli RM. Absence of correlation between infantile hypotonia and foramen magnum size in achondroplasia. Am J Med Genet. 2001;101:40–5.

33. Lutter LD, Longstein JE, Winter RB, Langer LO. Anatomy of the achondroplastic lumbar canal. Clin Orthop Relat Res. 1977;126:139–42.
34. Jeong ST, Song HR, Keny SM, Telang SS, Suh SW, Hong SJ. MRI study of the lumbar spine in achondroplasia. A morphometric analysis for the evaluation of stenosis of the canal. J Bone Joint Surg Br. 2006;88:1192–6.
35. Torley D, Bellus GA, Munro CS. Genes, growth factors and acanthosis nigricans. Br J Dermatol. 2002;147:1096–101.
36. Sabatino G, Di Rocco F, Zampino G, Tamburrini G, Caldarelli M, Di Rocco C. Muenke syndrome. Childs Nerv Syst. 2004;20:297–301.
37. Savarirayan R, Rossiter JP, Hoover-Fong JE, Irving M, Bompadre V, Goldberg MJ, et al. Best practice guidelines regarding prenatal evaluation and delivery of patients with skeletal dysplasia. Am J Obstet Gynecol. 2018;219:545–62.
38. White KK, Bompadre V, Goldberg MJ, Bober MB, Cho TJ, Hoover-Fong JE, et al. Best practices in peri-operative management of patients with skeletal dysplasias. Am J Med Genet A. 2017;173:2584–95.
39. Theroux MC, Nerker T, Ditro C, Mackenzie WG. Anesthetic care and perioperative complications of children with Morquio syndrome. Paediatr Anaesth. 2012;22:901–7.
40. Armstrong J, John J, Karsli C. A comparison between the GlideScope video laryngoscope and direct laryngoscope in paediatric patients with difficult airways—a pilot study. Anaesthesia. 2010;65:353–7.
41. Aziz MF, Healy D, Kheterpal S, Fu RF, Dillman D, Brambrink AM. Routine clinical practice effectiveness of the Glidescope in difficult airway management: an analysis of 2,004 Glidescope intubations, complications, and failures from two institutions. Anesthesiology. 2011;114:34–41.
42. Inada T, Fujise K, Tachibana K, Shingu K. Orotracheal intubation through the laryngeal mask airway in paediatric patients with Treacher-Collins syndrome. Paediatr Anaesth. 1995;5:129–32.
43. Theroux MC. Tracheal intubation for awake versus paralyzed infants with pyloric stenosis. Anesth Analg. 1998;87:1455; author reply 1456–7
44. Lattanzi DR, Harger JH. Achondroplasia and pregnancy. J Reprod Med. 1982;27:363–6.
45. Wardall GJ, Frame WT. Extradural anaesthesia for caesarean section in achondroplasia. Br J Anaesth. 1990;64:367–70.
46. Cohen SE. Anesthesia for cesarean section in achondroplastic dwarfs. Anesthesiology. 1980;52:264–6.
47. Morrow MJ, Black IH. Epidural anaesthesia for caesarean section in an achondroplastic dwarf. Br J Anaesth. 1998;81:619–21.
48. DeRenzo JS, Vallejo MC, Ramanathan S. Failed regional anesthesia with reduced spinal bupivacaine dosage in a parturient with achondroplasia presenting for urgent cesarean section. Int J Obstet Anesth. 2005;14:175–8.
49. Dubiel L, Scott GA, Agaram R, McGrady E, Duncan A, Litchfield KN. Achondroplasia: anaesthetic challenges for caesarean section. Int J Obstet Anesth. 2014;23:274–8.

Spondyloepiphyseal Dysplasia

6

Judith Jones, Lynda Arai, Eric Stickles,
and Lauren W. Averill

6.1 Introduction

Spondyloepiphyseal dysplasia (SED), or spondyloepiphyseal dysplasia congenita (SEDc), is an autosomal-dominant skeletal dysplasia described in 1966 by Spranger and Wiedemann [1]. It is a short-trunk skeletal dysplasia primarily affecting the vertebrae and the proximal epiphyses of long bones. The disease has a prevalence of 3.4/1,000,000, usually presenting at birth or in early childhood [2]. Spondyloepiphyseal dysplasia is characterized by delayed epiphyseal ossification. In this chapter, we use the abbreviation SEDc to refer to this skeletal dysplasia.

Infants with SEDc may be stillborn, born prematurely, or die shortly after birth from respiratory failure due to hypoventilation. The infants are short in stature because ossification may be absent or reduced in the vertebrae. The long bones and pubic bones also are dysplastic and become more so as the patient ages.

J. Jones (✉) · L. Arai
Department of Anesthesiology and Perioperative Medicine, Nemours Children's Hospital, Wilmington, Delaware, USA

Anesthesiology and Pediatrics, Sidney Kimmel Medical College, Thomas Jefferson University, Philadelphia, Pennsylvania, USA
e-mail: Judith.Jones@nemours.org; larai@nemours.org

E. Stickles
Department of Anesthesiology and Perioperative Medicine, Nemours Children's Hospital, Wilmington, Delaware, USA
e-mail: Eric.Stickles@nemours.org

L. W. Averill
Radiology and Pediatrics, Sidney Kimmel Medical College, Thomas Jefferson University, Philadelphia, Pennsylvania, USA

Department of Radiology, Nemours Children's Hospital, Wilmington, Delaware, USA
e-mail: Lauren.averill@nemours.org

M. C. Theroux (ed.), *Anesthesia Care of Patients with Skeletal Dysplasia*,
https://doi.org/10.1007/978-3-031-92501-6_6

Atlantoaxial instability occurs frequently in SEDc due to abnormalities in the cervical vertebrae. Knowledge of the natural history of this disease is critical for airway management since these children present for repeated anesthetic management during childhood and beyond. Other co-morbidities develop because bone abnormalities lead to structural and functional challenges, including restrictive lung disease, hypoventilation, and pulmonary hypertension. In rare instances, more mature patients with SEDc may present with tortuosity of the trachea; we have reported on such a patient [3]. Abnormal bone development may lead to coxa vara, genu valgum, pes planus, and dysplastic vertebral bodies. There may be platyspondyly or wedging, leading to significant kyphosis, scoliosis, and exaggerated lumbar lordosis of the spinal column. This further exacerbates the short stature seen in these patients.

In any patient suspected of SEDc, the diagnosis is based on history and physical examination, followed by initial radiological examination. Genetic testing is increasingly employed, and diagnosis is now largely co-dependent on genetic testing. Diagnosis is crucial for this patient population to ensure close monitoring and appropriate medical or surgical intervention as needed throughout the patient's life.

6.2 Genetics

Spondyloepiphyseal dysplasia congenita is an autosomal-dominant disorder that results from a mutation in the gene encoding for type II collagen, *COL2A1* [4]. The *COL2A1* gene is located on chromosome 12q13.1-q13.2, and it consists of 54 exons [2]. COL2A1, a large protein, is the major component of hyaline cartilage extracellular matrix, the nucleus pulposus of the intervertebral discs, the vitreous humor of the eye, and the structures of the inner ear. COL2A1 molecules have three identical alpha-1-polypeptide chains of 1060 amino acid residues each. There is a large triple-helical region, along with short, non-helical peptides. The COL2A1 molecules self-assemble into fibrils along with other macromolecules to form the extracellular scaffolding of cartilage critical for the structure and function of the cartilage. These fibrils crosslink to form mature COL2A1 fibers. Any molecular defects occurring in the *COL2A1* gene can cause a wide variety of rare, mostly autosomal-dominant conditions, called type II collagenopathies. To date, over 400 mutations of the *COL2A1* gene have been described in public databases and the literature, with 329 pathogenic variants and 153 variants of uncertain significance. These type II collagenopathies include SEDc, with the clinical features showing varying degrees of severity and complexity of co-morbid conditions. Over 100 *COL2A1* mutations have been identified in SEDc. Recently, a recessive pattern has also been demonstrated [2].

6.2.1 Phenotype

Phenotypes in SEDc vary widely. Presentation in infancy may occur as a stillborn or premature birth, as death shortly after delivery from respiratory failure due to hypoventilation, or as individuals who can mature beyond infancy. Variation in phenotypic expression is likely caused by gene polymorphisms and environmental

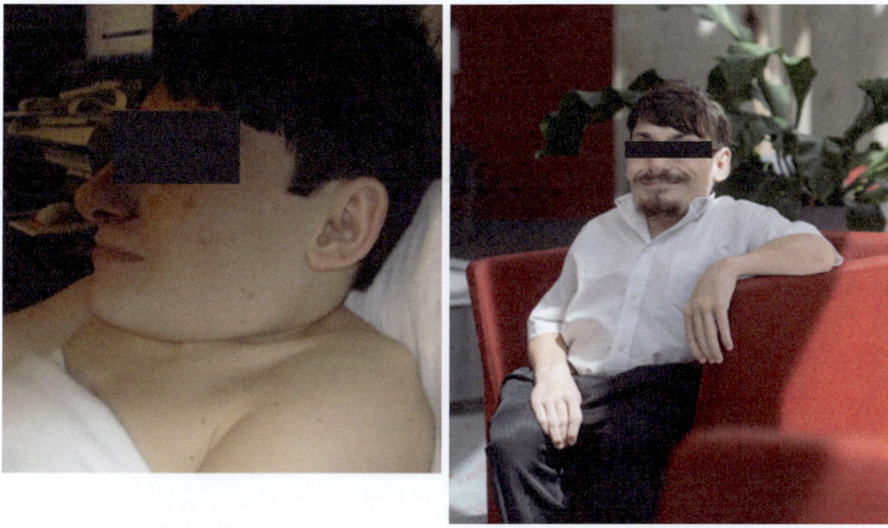

Fig. 6.1 Photographs of a male patient with SEDc. Photographs show the patient at 16 and 24 years of age. Note the short neck, almost not visible in the lateral profile. The thoracic cage is also shortened

factors [2]. Medical needs vary from one patient to another. The defect in the *COL2A1* gene leads to delayed ossification with bony effects, mainly impacting the respiratory system, the vertebral column, and the limbs, causing angular deformities of joints with ambulatory challenges. Figure 6.1 shows the features of SEDc in a young male patient.

Prenatal ultrasound of SEDc shows shortening of the long bones to <5th percentile with the fetus displaying flat facies. Postnatal measurements reveal infants who are short with mean length of 45 cm at term and flattened vertebrae (platyspondyly). Ossification may be absent or reduced, especially in the cervical and sacral vertebrae. As the infant with SEDc develops, atlantoaxial instability occurs frequently due to odontoid hypoplasia, os odontoideum (Fig. 6.2), ligamentous laxity, or hypoplasia of the body of the first cervical vertebra. This may lead to cervical spinal cord compression if the space available for the spinal cord (SAC) is critically low. Figure 6.2 shows narrowing of the spinal canal due to cervical instability.

C1–C2 instability may progress to cervical subluxation or dislocation with possible cervical myelopathy due to spinal cord injury. Quadriplegia and respiratory failure may occur as a consequence. It is especially important to bear these issues in mind during airway management since these children present repeatedly for anesthetic care for diagnostic imaging and orthopedic procedures during childhood and beyond. In fact, consensus-based guidelines were published in 2017 indicating that all skeletal dysplasia patients with significant spinal deformity, including SEDc patients, should have clinical and/or imaging evaluations done to detect and treat abnormalities of the spine to prevent or minimize spinal cord injury [5].

Infants with SEDc may display ovoid or pear-shaped vertebrae and later develop platyspondyly and irregular end-plates in the childhood years.

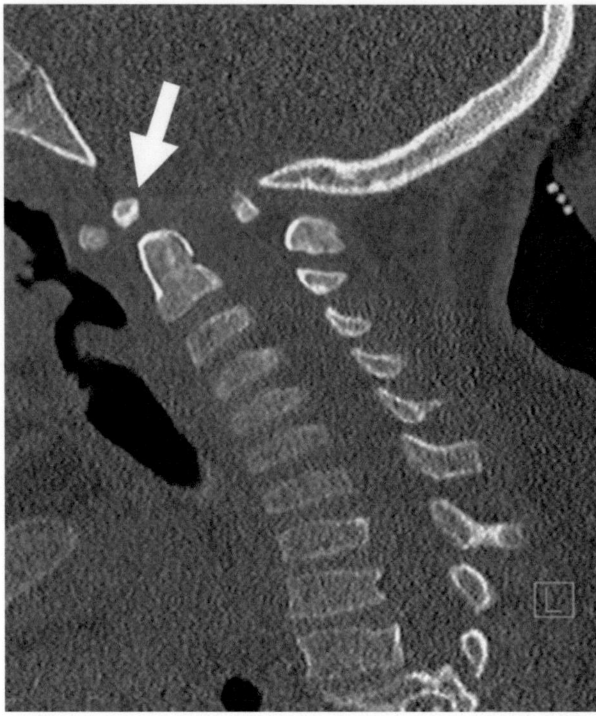

Fig. 6.2 A computed tomography (CT) scan of an 11-year-old female with SEDc and upper cervical instability. Sagittal reconstruction from a cervical spine CT shows os odontoideum (arrow). The ring of C1 is seen in cross-section and is displaced anteriorly compared to the body of C2. Likewise, the posterior ring of C1 is displaced anteriorly, encroaching on the spinal canal. Abnormal C1–C2 motion was seen on flexion-extension cervical spine magnetic resonance imaging (MRI) (not shown), indicating atlantoaxial instability commonly seen in SEDc. The patient underwent C1 decompression and C1–C2 spinal fusion 3 days after the CT scan

Platyspondyly leads to significant short stature, which may not be evident until age 2–3 years. Thoracolumbar kyphoscoliosis may develop over time. Significant upper thoracic kyphosis can lead to spinal cord injury if left untreated, especially during prolonged procedures when sedation or anesthesia is required for patient immobility. There have been reports of patients who sustained spinal cord injury—documented clinically and shown on MRI—after prolonged periods in the supine position under anesthesia. This has led to the recommendation that all patients with significant upper thoracic kyphosis have neurophysiology monitoring for procedures longer than an hour that require sedation or general anesthesia [5, 6].

Ossification may be absent or delayed in the pubic bones and in the epiphyses of the long bones with delay in the measured bone age. This delay in ossification may lead to the development of abnormalities such as absent femoral heads due to flattened epiphyses, coxa vara, and genu valgum. The roof of the acetabulum may be more horizontal compared to its normal orientation. Additionally, the iliac bones may be shorter in length without the normal flaring of the wings of these bones. There also may be pes planus deformity or contractures at the elbows. Shortened limbs are typical. These abnormalities often lead to pain or difficulties with patient ambulation.

Other co-morbidities also develop due to bone abnormalities leading to structural and functional challenges for patients with SEDc. These include cleft palate, pectus carinatum, and barrel chest. Abnormalities outside of the musculoskeletal system also occur, such as mitral valve prolapse, sensorineural hearing loss, myopia, and retinal detachment.

6.3 Radiographic Findings

Although diagnostic evaluation of a patient with suspected SEDc always begins with a thorough medical history and physical examination, imaging studies aid confirmation of a diagnosis. A complete skeletal survey is usually the first step in imaging and includes the following: anteroposterior (AP) and lateral skull films; cervical spine film with AP and lateral views in flexion, neutral position, and extension; posteroanterior (PA) views of the wrist and hand; AP and lateral projections of the elbows, hips, and knees; AP and lateral views of the thoracolumbar vertebrae; and AP film of the lumbar and sacral regions [7, 8].

A generalized delay in the formation of ossification centers is a hallmark of SEDc [9]. The skeletal survey will demonstrate the absence of epiphyseal centers in the distal femur, proximal tibia, os pubis, calcaneus, and talus, which should all be present at birth. When epiphyses are visible later in life, they appear flattened and irregular in shape. Oval, trapezoid, or pear-shaped vertebrae are often present on radiographic imaging of the spine. Once patients reach adolescence, end-plate irregularities and narrowed intervertebral disk spaces become obvious. Most markedly, gross ventral hypoplasia may be present at the thoracolumbar junction. Lumbar lordosis is usually exaggerated, and progressive kyphoscoliosis may develop in late childhood.

Figure 6.3 shows the plain radiographs of a patient with SEDc who displays exaggerated lumbosacral lordosis.

Atlantoaxial instability is another complication occurring secondary to odontoid hypoplasia and endows a greatly increased risk of cervical myelopathy (Fig. 6.2) [10]. Other features of SEDc frequently noted on radiographic imaging include large and dolichocephalic skull and rhizomelic shortening of extremities, more commonly involving the lower limbs [10].

MRI is another useful imaging modality used to diagnose and guide treatment in patients with SEDc. As C1–C2 instability is common, an MRI can be used to delineate cord compression and cervical myelopathy. This is important information to obtain, if possible, prior to any planned instrumentation of the airway or anesthetic procedure [11]. Figure 6.4 shows the MRI of a patient with cervical spinal cord compression. The existence of any spinal abnormalities is important information to have prior to an anesthetic as this may dictate the anesthetic management as well as

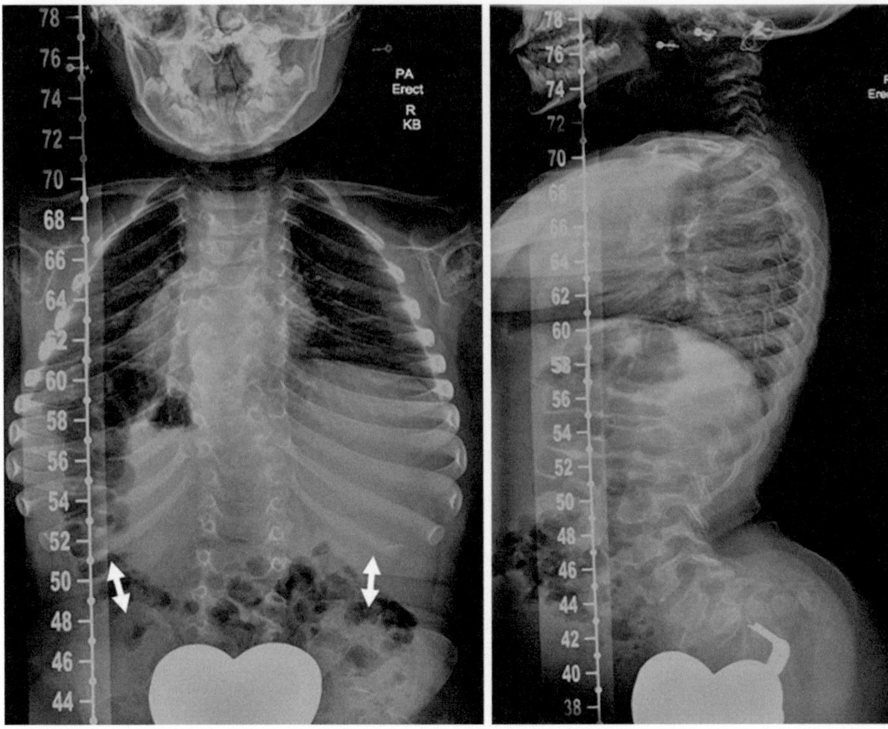

Fig. 6.3 Frontal and lateral spine radiographs of an 8-year-old girl with SEDc show platyspondyly, especially in the cervical and thoracic spine, as well as exaggerated lumbosacral lordosis. Note the short trunk with close position of the rib cage to the iliac crests (double arrows). She has mild scoliosis at this age. The ruler running vertically on the edge of the image is used to align multiple images stitched together into a composite image of the spine

the need for neuromonitoring and maintaining an adequate mean arterial pressure to optimize cord perfusion.

Severe kyphoscoliosis and exaggerated lumbar lordosis can be easily visualized on an MRI of the lumbar spine. Figure 6.5 shows the MRI images of a patient with severe kyphoscoliosis and exaggerated lordosis.

Finally, vision and hearing tests are performed commonly as part of a patient's diagnostic evaluation due to the high incidence of co-morbid conditions affecting these organ systems.

In families with a high risk of SEDc, prenatal diagnosis may be suggested by ultrasound findings of delayed ossification of the pubis and epiphysis of the femur and humerus. Sonolucency secondary to delayed ossification of the vertebral bodies may be seen. They may appear flattened or small. Head size will appear normal on ultrasound imaging [12].

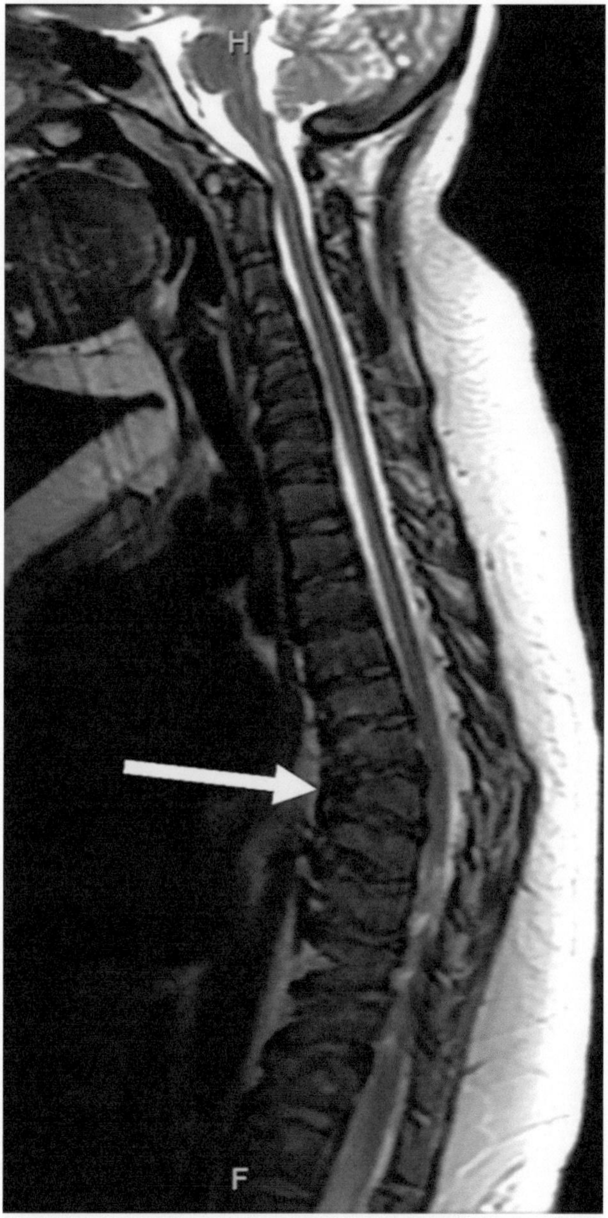

Fig. 6.4 MRI of a14-year-old female with SEDc. Sagittal T2-weighted scout image from an MRI of the spine shows marked platyspondyly and end-plate irregularity, especially in the thoracic spine, with anterior wedging of T7 (arrow) and associated kyphosis. Note cord compression at the craniocervical junction

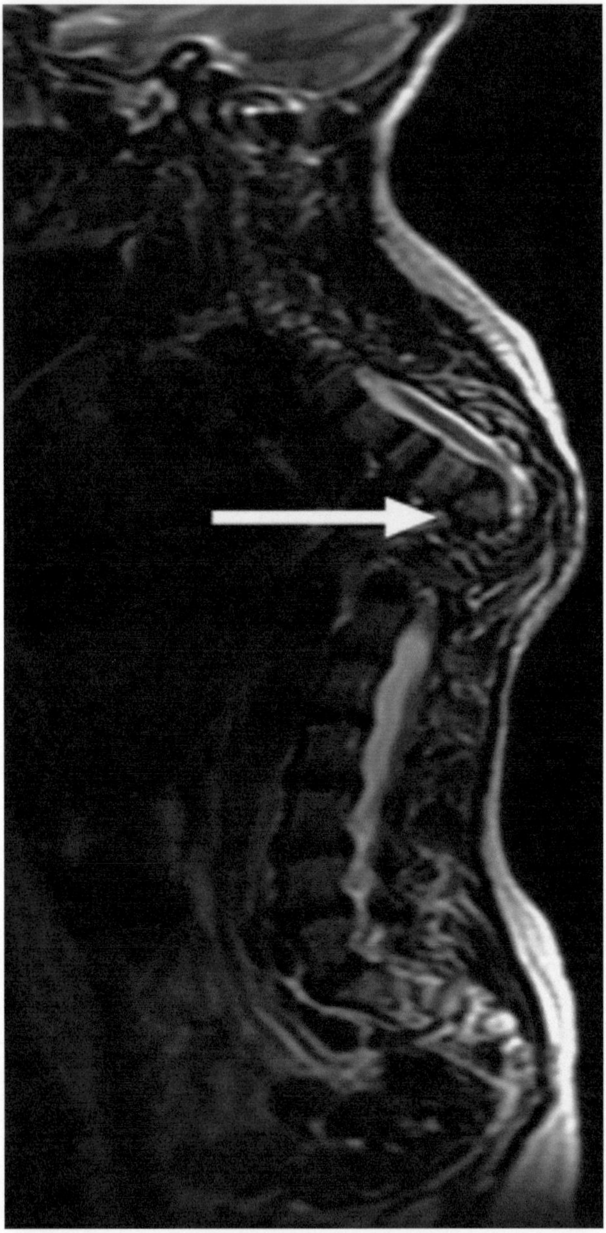

Fig. 6.5 An MRI of a 28-year-old woman with SEDc. Sagittal T2-weighted scout image from a spine MRI shows severe kyphoscoliosis (arrow). The entire spine cannot be displayed on a single sagittal image due to the severe curvature. Exaggerated lordosis of the lumbar spine also is visible on this MRI image

6.4 Co-morbid Conditions

Genetic mutations leading to SEDc can affect collagen and connective tissues throughout the body and are not limited to the spine and epiphyseal centers. Ocular anomalies are the most associated conditions, and the association of SEDc with myopia, retinal detachment, and deafness was first described in 1969 [13]. The incidence of myopia and retinal detachment are estimated to be 45% and 12%, respectively in SEDc patients [14]. Other common ocular manifestations include buphthalmos (one or both eyeballs are larger than normal), secondary glaucoma, and strabismus [14]. One important ocular finding is that the corneas remain clear in patients with SEDc. Craniofacial deformities are common in patients with SEDc. These include cleft palate, flat face, and hypertelorism (wide-set eyes). A pectus carinatum deformity may be present in combination with a barrel-shaped chest and may lead to respiratory abnormalities, such as the development of obstructive sleep apnea (OSA). The development of severe kyphoscoliosis in certain patients can lead to restrictive lung disease and, ultimately, cor pulmonale in the later stages of adolescence. Although typically found in the milder form of SEDc (known as SED tarda), in very rare instances, SEDc is associated with nephrotic syndrome [15].

At the time of writing, there are 84 patients in the cohort of SEDc patients cared for at Nemours Children's Hospital Delaware. The patients range in age from 16 months to 34 years. Of these patients, 24 (29%) had radiological features of cervical spine instability. As a result, most patients were intubated using video laryngoscopy. Forty-two patients were intubated using the Glidescope (Verathon; Washington, USA) and four patients using the C-MAC system (Karl Storz SE & Company; Tuttlingen, Germany). Three patients required two-hand, two-person bag-mask ventilation, while five patients required tongue retraction to facilitate blade placement during laryngoscopy. Three patients (4%) had tracheostomy placement in early childhood due to significant airway obstruction. Twelve patients (14%) had cleft palate, and four patients (5%) had a diagnosis of Pierre Robin sequence. Eight patients (10%) had OSA.

Forty-three patients (51%) had procedures for lower extremity abnormalities, eleven patients (13%) had posterior spinal fusion, and five patients (6%) had cervical spine decompression and/or fusion. The most common orthopedic abnormalities were scoliosis (24%), hip dysplasia (20%), and coxa vara (18%). Genu valgum was present in 11% of patients, and 10% of patients had hearing loss.

6.5 Anesthesia and Postoperative Care

Children with SEDc often present for surgery to correct limb or spine deformities as well as for other abnormalities. Significantly greater risk is involved in caring for these patients due to multiple factors, including airway abnormalities and other co-morbidities previously mentioned. It is imperative that the clinician is knowledgeable about the disorder and the individual patient's characteristics. These patients should be cared for in facilities that are aware of and can manage potential life-threatening complications [16].

A thorough preoperative evaluation of a patient with SEDc should be performed, including evaluation of the airway, cardiopulmonary function, and complete neurologic history and examination. The history should focus on pulmonary co-morbidities, including upper airway abnormalities and obstruction, trachea-bronchial abnormalities, and pulmonary impairment due to significant thoracic dysplasia. A sleep study prior to major surgery should be considered in patients who have symptoms suggestive of significant OSA. Pulmonary function studies may be helpful in evaluating pulmonary compromise due to thoracic cage abnormalities that typically result in restrictive lung disease. Patients with chronic and severe restrictive lung disease also may develop chronic hypoxia and pulmonary hypertension. If this is suspected, then a complete cardiac evaluation should be performed by a cardiologist. A thorough neurologic assessment should focus on signs and symptoms of myelopathy, including unsteady gait, extremity weakness, and bowel or bladder incontinence [16]. The neurological examination also should look for signs and symptoms of cervical instability, which include hypotonia, sleep apnea, and respiratory insufficiency. Cervical cord compression, in addition to thoracic dysplasia, can contribute to respiratory insufficiency. Therefore, children with pulmonary abnormalities, especially respiratory muscle weakness, should be evaluated fully for cervical instability [17].

One of the most concerning aspects of anesthetic management in SEDc patients is the potential for difficult airway management. Our patients are often intubated using a video laryngoscope due to cervical spine abnormalities, and, therefore, a true "difficult intubation" is hard to tease out. However, there were nine patients who were "difficult airway" as noted in the chart. The reasons for a difficult airway can be multi-factorial and related to the physical characteristics of short neck, micrognathia and laryngotracheal stenosis, as well as to those of cervical abnormalities, particularly odontoid hypoplasia and atlantoaxial instability. Airway management may be even more challenging if the patient had a previous cervical fusion with resultant cervical immobility. A high proportion of patients with SEDc are at risk for atlantoaxial instability, and the anesthesiologist should be on high alert for this [17]. The cervical spine should be fully evaluated prior to anesthetic care with at least cervical spine plain films, including flexion and extension views, when determining instability. However, plain films may be difficult to interpret and provide no information about the underlying spinal cord. Flexion-extension MRI has been shown to be safe and can provide a more thorough evaluation of any spinal cord compression, if present [18]. If the patient has cervical instability but is presenting for a procedure not involving the cervical spine, an orthopedic specialist and neurosurgeon should be consulted for evaluation. If the cervical instability is significant, then the discussion may need to focus on correction of the cervical spine instability prior to the patient having any other procedures, as this can potentially progress to subluxation and quadriparesis [19].

The anesthesiologist should have multiple tools available to intubate a patient with SEDc if a conventional intubation is not successful [20]. A full history of previous intubations should be obtained with the most recent anesthetic records being most pertinent. If there is significant concern for intubation, it may be advisable to have an otolaryngologist available to assist if necessary, and this practice will vary between different institutions and different countries. Any reports of previous difficulty with intubation at an outside institution should be investigated, and the anesthesiologist should determine the devices and methods used to achieve intubation.

Another concerning aspect of intraoperative management of a patient with SEDc is the risk of spinal cord injury during prolonged procedures in the supine position. There are reports of thoracic spinal cord infarct after prolonged anesthesia in patients with skeletal dysplasia who were not having spine procedures [19, 21]. In a report by Pruszczynski et al., the authors performed a retrospective review of children having lower extremity surgery and found 2 of 52 patients who had developed paraplegia following surgery [19]. Both patients had prolonged operative times and significant thoracic kyphosis. One of the patients had SEDc, and the other had Morquio 1-H. The patient with SEDc recovered motor function after 5 months. Both patients in this report had intermittent periods of low mean arterial pressure (MAP), which was felt to be contributory (in combination with positioning and severe kyphosis) to the spinal cord ischemia. Another case report describes a patient with Morquio A and spinal stenosis who developed paraplegia following lower extremity surgery [21]. This patient had a lumbar epidural placed while awake, which was utilized in combination with general anesthesia for the procedure. The patient had intermittent MAP below baseline during the procedure, and since the epidural was being utilized, the paraplegia in the recovery period was not immediately recognized. The authors, Drummond et al. [21] postulated that the significant spinal stenosis in combination with suboptimal MAP contributed to the spinal cord ischemia. Neither of these patients had neuromonitoring during the procedure. Neuromonitoring could have alerted the practitioners to spinal cord compromise, resulting in adjustment of patient position and elevation of MAPs during the procedure. An MRI of the entire spine is an important tool for patients with neurologic signs or symptoms and can be useful in assessing spinal cord compromise preoperatively. However, its utility is not clear for surveillance purposes in asymptomatic patients [5, 16].

Neuromonitoring should be available for all procedures longer than 1 h in patients considered to have "at risk" spine abnormalities or spinal cord compromise. This is based on a consensus statement from skeletal dysplasia experts that defined "spine-at-risk" findings to include spinal cord impingement from atlantoaxial instability, cervical stenosis, foramen magnum stenosis, cervical kyphosis, cervicothoracic kyphosis, thoracic spinal stenosis, thoracolumbar kyphosis, syrinx, and spinal cord signal change on MRI [5, 16]. If neuromonitoring is utilized, inhalational agents should be avoided, and total intravenous anesthesia should be provided with continuous infusion of other agents, such as propofol or ketamine, in addition to opiates provided through continuous infusion or intermittent boluses. These patients may require higher MAPs to maintain optimal spinal cord perfusion. This should be a consideration and should be discussed with the surgeon as well as the neurophysiologist.

To allow for adequate assessment of positioning, neuromonitoring should be initiated with baselines obtained prior to positioning the patient. A comparison of the somatosensory and motor evoked responses prior to and after positioning can be made and adjustments performed to maximize the neurologic response. Patients with SEDc and secondary scoliosis and hyperkyphosis were shown to be more likely to have loss of neuromonitoring signals with positional changes when presenting for spine surgery [22]. In one study, children were monitored in the supine position and then placed prone. Six patients showed permanent loss of neuromonitoring signals, and four of the six experienced the change with prone positioning. The commonality for all six patients was the significant degree of kyphosis. Another published case report describes a patient with mucopolysaccharidosis (MPS), also a skeletal dysplasia, who developed paraplegia despite no change in neuromonitoring response during the procedure. In this case, the baseline responses were abnormal but were not obtained until after positioning [23]. The abnormal baseline was attributed to the child's MPS, and, since there was not a pre-positional response to compare it to, the team was not alerted to the possibility of spinal cord compromise. This highlights the importance of neuromonitoring throughout the induction and positioning period. If the patient develops significant changes seen on neuromonitoring during the induction period, these may be ameliorated by repositioning the patient in the lateral position as well as increasing the MAP with vasoactive agents if necessary.

Many younger patients with SEDc can proceed with inhalation induction of general anesthesia; although, it is imperative to establish intravenous access early in the induction process. If neuromonitoring is to be used, the inhalational agents should be discontinued as soon as possible once intravenous access is established and prior to achieving baseline evoked potentials. The head and neck should be maintained in a neutral position, which is achieved best by slight neck extension facilitated by a small shoulder roll to align the pinna of the ear with the clavicle [24]. It is helpful to have an assistant retract the tongue with ring forceps during laryngoscopy, thereby effectively increasing space in the oral cavity for intubation. An awake, sedated, fiberoptic intubation should be considered and neuromonitoring utilized throughout intubation if there is significant cervical instability.

Epidural analgesia has been utilized in patients with SEDc, although this is no longer considered appropriate and is not advisable. Two case reports describe successful epidural anesthesia in adult patients with SEDc presenting for cesarean section [25, 26]. In another case report, also for an adult patient presenting for cesarean section, the patient received general anesthesia after awake fiberoptic intubation because of previous difficulty with epidural placement and known difficult intubation [27]. However, as previously described, there are many reasons to avoid epidural anesthesia and analgesia in this patient population. If neuromonitoring is utilized, then an epidural anesthetic would interfere with monitoring. The placement of an epidural catheter could be challenging due to significant kyphoscoliosis and a shallow epidural space [26]. In addition, it would be difficult to evaluate the patient postoperatively for any signs of neurologic compromise if an epidural is actively utilized with local anesthetic. Epidural anesthesia also can significantly lower MAP, which would have a deleterious effect on spinal cord blood flow and contribute to spinal cord compromise. Alternatively, regional techniques such as peripheral nerve blocks have been used successfully and should be considered for major limb procedures to provide postoperative analgesia [28]. The ability to monitor motor function postoperatively would still be a concern and therefore requires discussion with the surgeon. If neuromonitoring is utilized intraoperatively, then the peripheral nerve blocks would need to be initiated at the conclusion of the procedure.

Patients with SEDc may present for joint replacement as their disease progresses due to significant pain and immobility. These patients often require joint replacement at an earlier age than the average population. A retrospective review of patients with skeletal dysplasia presenting for joint replacement investigated risk factors for surgery and provided care recommendations for this patient population [29]. They identified that surgical complications were highest in patients with cardiopulmonary co-morbidities and recommended thorough preoperative evaluation to identify other co-morbidities as well as comprehensive evaluation of spine abnormalities. In this study of 29 individuals who had a total of 64 joint replacements, SEDc was the most common skeletal dysplasia.

Postoperative pain management may be challenging for patients with SEDc who have chronic pain or major limb or spine surgery. As stated, regional techniques can be considered but may be more difficult to perform in this population due to anatomical abnormalities. Epidural analgesia is not recommended, as previously stated, due to concerns for spinal cord compromise and the inability to evaluate spinal cord function effectively in the postoperative period. Multimodal analgesics should be employed for postoperative analgesia, including non-steroidal anti-inflammatory agents, acetaminophen, and short-acting opiates. Diazepam or other benzodiazepines can be useful to treat painful muscle spasm after extensive limb or spinal surgery. Gabapentin and clonidine also can be employed as adjuncts and may be beneficial in reducing opiate requirements. Patients with known OSA should receive reduced amounts of opiates and benzodiazepines due to concern for respiratory depression. Postoperative admission to an intensive care unit setting should be considered for patients with severe OSA or other significant co-morbidities requiring closer observation and monitoring.

6.6 Case Illustration

A 22-month-old female was delivered at 37 + 2 weeks gestation via cesarean section. The postnatal course was complicated by a stay in the neonatal intensive care unit for low oxygen saturation in the first 24 h of life.

After birth, the patient was noted to have feeding difficulties, a submucosal cleft palate, and a history of snoring and apnea. Upper cervical instability and stenosis of the foramen magnum were visualized on an MRI conducted in the neonatal period. She was diagnosed with SEDc on genetic testing. At the age of 5 months, the patient had an uneventful C1 laminectomy and decompression of the spinal cord due to spinal cord compression. She wore a soft cervical collar for cervical spine support due to cervical spine instability, seen in Fig. 6.6.

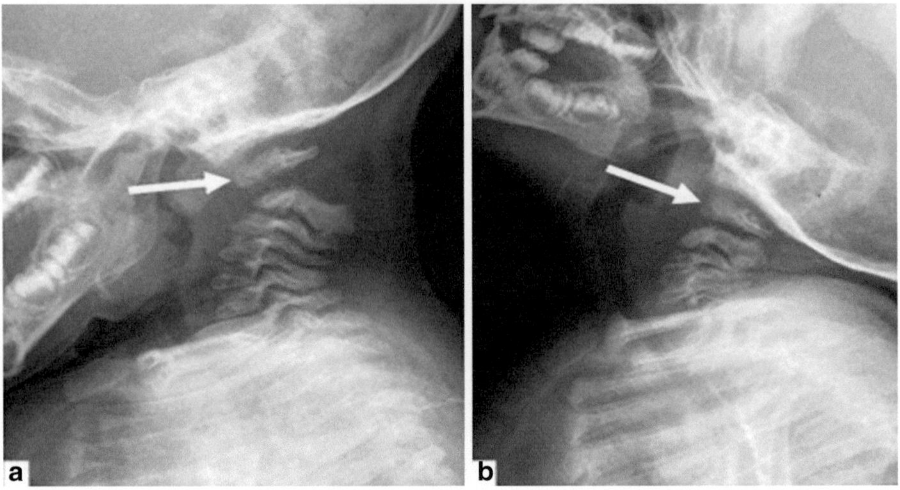

Fig. 6.6 Lateral cervical spine radiographs with neck flexion (**a**) and extension (**b**) performed on 16-month-old female with SEDc show atlantoaxial instability with significant motion and angulation of C1 (arrows) in relation to C2

At age 22 months, the patient presented for halo application and posterior cervical fusion of the occiput to C2 due to significant instability at the atlanto-occipital junction seen on cervical spine X-ray, with narrowing of the spinal canal at C1–C2 (Figs. 6.7 and 6.8). Thoracolumbar kyphosis, which was noted on a thoracolumbar spine X-ray, improved from 71° of kyphosis at age 16 months to 53° of kyphosis at the same spinal cord level at age 22 months.

The patient had an uneventful mask induction with easy mask ventilation and intravenous catheter placement. She was easily intubated using a 3.5-mm oral endotracheal tube and video laryngoscopy. The neck was maintained in neutral position for airway management. Baseline readings were obtained for motor and somatosensory evoked potentials. The halo was applied, and the patient was turned carefully to the prone position with care to maintain neutrality of the cervical spine during the change in position.

There was an alert indicated by the neurophysiologist due to a sudden absence of motor and somatosensory evoked potentials from bilateral lower extremities. The patient was immediately returned to the supine position, and the evoked potentials returned in both lower extremities. After a short period of time, it was noted that the motor evoked potential amplitude was reduced in the right lower extremity, so the patient was placed in lateral decubitus position. The surgical and anesthesia teams discussed the possible source of disruption of the evoked potentials. They thought it likely that there was spinal cord dysfunction because of compression at the thoracolumbar level due to significant thoracic kyphosis. The decision was made not to proceed with posterior upper cervical fusion and decompression; instead the patient was evaluated for spinal cord compromise from her known thoracolumbar kyphosis.

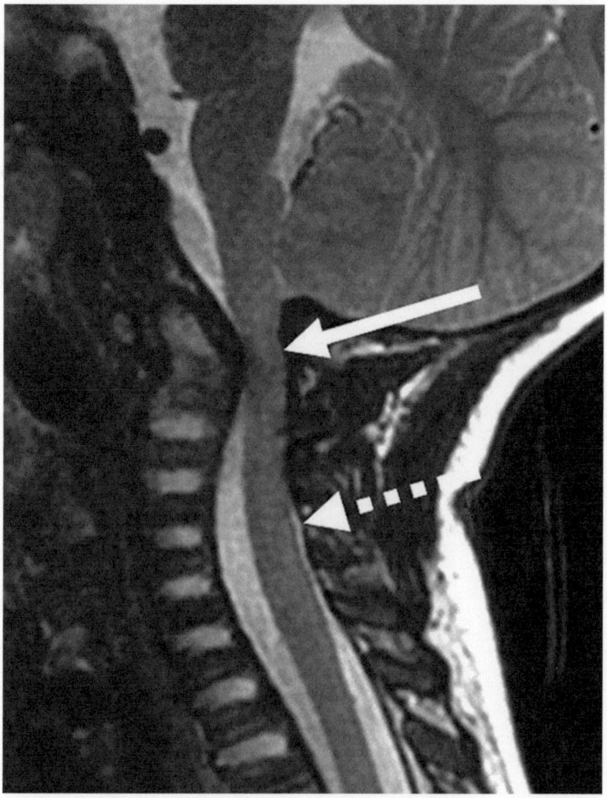

Fig. 6.7 Sagittal T2-weighted MRI of the cervical spine performed at 22 months shows spinal canal narrowing at C1–C2 (arrow) and normal caliber in the mid-cervical spine (dashed arrow)

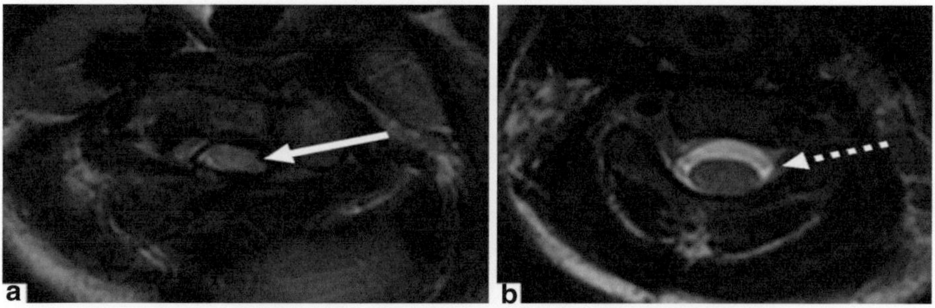

Fig. 6.8 Axial T2-weighted image at C1–C2 (**a**) shows complete effacement of cerebrospinal fluid (CSF) and mildly increased signal in the spinal cord (arrow). Axial T2-weighted image in the mid-cervical spine (**b**) shows the normal appearance of the spinal cord and CSF for comparison (dashed arrow) in this 22-month-old patient

A wake-up test was done in the operating room, and the patient moved all extremities well. The anesthesia team transported the patient to the MRI scanner, still intubated, for imaging of the entire spine. No acute changes were noted on the MRI scan, and the patient had an uneventful emergence from the anesthetic. An MRI of the patient's thoracic and lumbar spine showed significant kyphosis (Fig. 6.9).

The patient was observed overnight and was discharged on postoperative day 1 in stable condition with a plan to return to the orthopedic clinic in 6 weeks. She continued to move all extremities with baseline strength at the time of discharge.

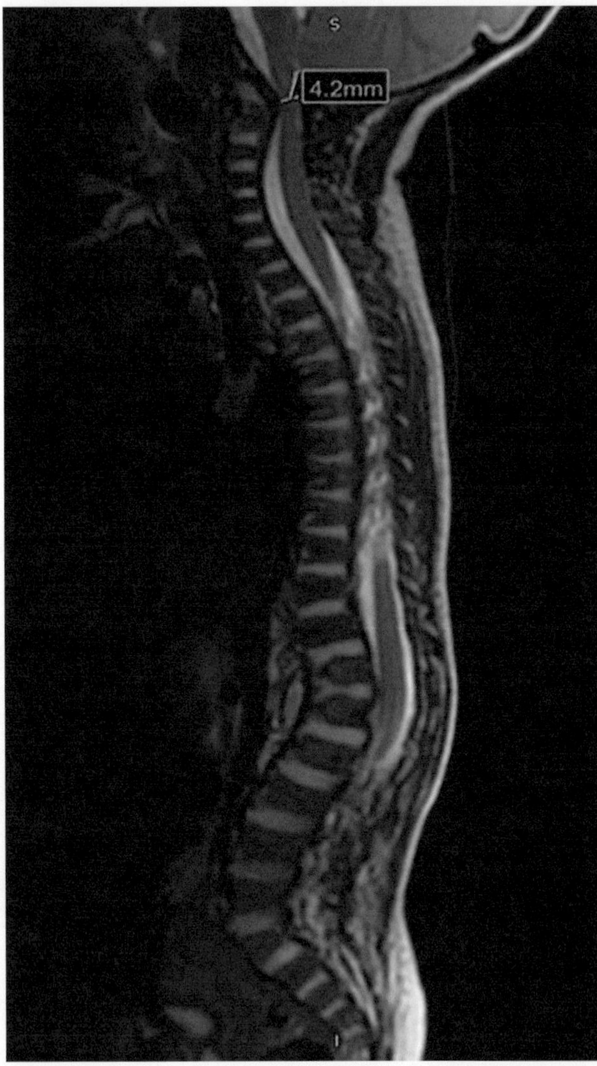

Fig. 6.9 Sagittal T2-weighted spine MRI showing significant focal kyphosis at the thoracolumbar junction

Five months later, the patient presented to the operating room for thoracolumbar fusion. Anesthesia was induced via inhalation induction, which was carried out smoothly in the supine position. There was easy mask ventilation with easy placement of an intravenous catheter and 3.5 cuffed endotracheal tube using video laryngoscopy. The cervical spine was maintained in neutral position during induction and endotracheal intubation. A propofol infusion was started, and the patient was turned to the left lateral position for placement of electrodes for somatosensory and motor evoked potential monitoring, an additional intravenous catheter, a radial artery catheter for invasive blood pressure monitoring and blood draws for laboratory investigations, and a Foley catheter for monitoring of the patient's urine output. The patient's head was placed in pins while the patient was in the lateral position. The lateral position was maintained to minimize the risk of spinal cord compromise, which occurred during the previous surgical attempt 5 months earlier.

The patient was then turned carefully to the prone position with monitoring of pre- and post-turn motor evoked potentials. No adverse events were noted, and the surgical procedure proceeded without any spinal cord compromise noted on neurophysiological monitoring. The patient had an unremarkable intraoperative and postoperative course and was discharged to home in stable condition. The patient is scheduled for cervical spine decompression and fusion at a later date.

References

1. Spranger J, Wiedeman HR. Dysplasia spondyloepiphysaria congenita. Helv Paediatr Acta. 1966;21:598–611.
2. Nenna R, Turchetti A, Mastrogiorgio G, Midulla F. COL2A1 gene mutations: mechanisms of spondyloepiphyseal dysplasia congenita. Appl Clin Genet. 2019;12:235–8.
3. Stefko J, Oommen A, Jones J, Arai L, DiCindio S, Theroux M. Severe multilevel tracheal stenosis with significant twisting in a patient with spondylometaphyseal dysplasia: a case report. Anaheim, CA: SPA 37th Annual Meeting; 2023.
4. Zhou T, Yang X, Chen Z, Zhou Y, Cao X, Zhao C, et al. A novel COL2A1 mutation causing spondyloepiphyseal dysplasia congenita in a Chinese family. J Clin Lab Anal. 2021;35:e23728.
5. White KK, Bompadre V, Goldberg MJ, Bober MB, Cho TJ, Hoover-Fong JE, et al. Best practices in peri-operative management of patients with skeletal dysplasias. Am J Med Genet A. 2017;173:2584–95.
6. Brothers J, DiCindio S, Thacker M, Mackenzie S, Campbell J, Mackenzie WG, Theroux MC. Recommendations to treat changes in evoked potentials in patients with skeletal dysplasia. Austin, TX: SPA/AAP Pediatric, Pediatric Anesthesiology Annual Meeting; 2023.
7. Shapiro F. Review of specific skeletal dysplasias: pathobiology, clinical and radiographic characteristics. In: Shapiro F, editor. Pediatric orthopedic deformities, vol. 1. Cham: Springer; 2016. p. 324–98.
8. Spranger JW, Langer LO Jr. Spondyloepiphyseal dysplasia congenita. Radiology. 1970;94:313–22.
9. Handa A, Grigelioniene G, Nishimura G. Radiologic features of type II and type XI collagenopathies. Radiographics. 2021;41:192–209.
10. Panda A, Gamanagatti S, Jana M, Gupta AK. Skeletal dysplasias: a radiographic approach and review of common non-lethal skeletal dysplasias. World J Radiol. 2014;6:808–25.
11. Redl G. Massive pyramidal tract signs after endotracheal intubation: a case report of spondyloepiphyseal dysplasia congenita. Anesthesiology. 1998;89:1262–4.

12. Dighe M, Fligner C, Cheng E, Warren B, Dubinsky T. Fetal skeletal dysplasia: an approach to diagnosis with illustrative cases. Radiographics. 2008;28:1061–77.
13. Fraser GR, Friedmann AI, Maroteaux P, Glen-Bott AM, Mittwoch U. Dysplasia spondyloepiphysaria congenita and related generalized skeletal dysplasias among children with severe visual handicaps. Arch Dis Child. 1969;44:490–8.
14. Terhal PA, Nievelstein RJ, Verver EJ, Topsakal V, van Dommelen P, Hoornaert K, et al. A study of the clinical and radiological features in a cohort of 93 patients with a COL2A1 mutation causing spondyloepiphyseal dysplasia congenita or a related phenotype. Am J Med Genet A. 2015;167A:461–75.
15. Lama G, Marrone N, Majorana M, Cirillo F, Salsano ME, Rinaldi MM. Spondyloepiphyseal dysplasia tarda and nephrotic syndrome in three siblings. Pediatr Nephrol. 1995;9:19–23.
16. White KK, Bober MB, Cho TJ, Goldberg MJ, Hoover-Fong J, Irving M, et al. Best practice guidelines for management of spinal disorders in skeletal dysplasia. Orphanet J Rare Dis. 2020;15:161.
17. Al Kaissi A, Ryabykh S, Pavlova OM, Ochirova P, Kenis V, Chehida FB, et al. The management of cervical spine abnormalities in children with spondyloepiphyseal dysplasia congenita: observational study. Medicine (Baltimore). 2019;98:e13780.
18. Mackenzie WG, Dhawale AA, Demczko MM, Ditro C, Rogers KJ, Bober MB, et al. Flexion-extension cervical spine MRI in children with skeletal dysplasia: is it safe and effective? J Pediatr Orthop. 2013;33:91–8.
19. Pruszczynski B, Mackenzie WG, Rogers K, White KK. Spinal cord injury after extremity surgery in children with thoracic kyphosis. Clin Orthop Relat Res. 2015;473:3315–20.
20. Apfelbaum JL, Hagberg CA, Connis RT, Abdelmalak BB, Agarkar M, Dutton RP, et al. 2022 American Society of Anesthesiologists practice guidelines for management of the difficult airway. Anesthesiology. 2022;136:31–81.
21. Drummond JC, Krane EJ, Tomatsu S, Theroux MC, Lee RR. Paraplegia after epidural-general anesthesia in a Morquio patient with moderate thoracic spinal stenosis. Can J Anaesth. 2015;62:45–9.
22. Rizkallah M, El Abiad R, Badr E, Ghanem I. Positional disappearance of motor evoked potentials is much more likely to occur in non-idiopathic scoliosis. J Child Orthop. 2019;13:206–12.
23. Tong CK, Chen JC, Cochrane DD. Spinal cord infarction remote from maximal compression in a patient with Morquio syndrome. J Neurosurg Pediatr. 2012;9:608–12.
24. Theroux MC, Lopez M, Olszewsky PJ, DiCindio S, Arai L, Ditro C, et al. Metatropic dysplasia-a skeletal dysplasia with challenging airway and other anesthetic concerns. Paediatr Anaesth. 2017;27:596–603.
25. de Boer HD, Hemelaar A, van Dongen R, Gielen MJ. Successful epidural anaesthesia for caesarean section in a patient with spondyloepiphyseal dysplasia. Br J Anaesth. 2001;86:133–4.
26. Rodney GE, Callander CC, Harmer M. Spondyloepiphyseal dysplasia congenita. Caesarean section under epidural anaesthesia. Anaesthesia. 1991;46:648–50.
27. Falzon D, Burns R, Theodosiou CA, Cooper ES. Awake fibreoptic intubation for caesarean section in a patient with spondyloepiphyseal dysplasia congenita. Int J Obstet Anesth. 2018;33:96–7.
28. Eiszner JR, Atanda A Jr, Rangavajjula A, Theroux M. A case series of peripheral nerve blocks in pediatrics and young adults with skeletal dysplasia. Paediatr Anaesth. 2016;26:553–6.
29. Raggio CL, Yonko EA, Khan SI, Carter EM, Citron KP, Bostrom MPG, et al. Joint replacements in individuals with skeletal dysplasias: one institution's experience and response to operative complications. J Arthroplast. 2020;35:1993–2001.

Kniest Dysplasia

7

Malgorzata Lutwin-Kawalec, Susmita Nesargi,
Alfred T. Dorsey, and Lauren W. Averill

7.1 Genetics

Kniest dysplasia (KD), also called metatropic dwarfism type II, pseudometatropic dysplasia, or Swiss-cheese cartilage dysplasia, is a type of collagenopathy caused by a mutation in the collagen type II *COL2A1* gene located on chromosome 12. This exon-skipping mutation causes a specific amino acid deletion at the c-terminus of the type II collagen helix between exons 12 and 24, shortening of collagen monomers, and disruption of the normal triple helix configuration, resulting in procollagen retention in chondrocyte endoplasmic reticulum [1–6]. Most cases are caused by a sporadic mutation in the offspring of parents with average height. The estimated incidence is close to 1:1,000,000, with males and females equally affected [7, 8]. The inheritance is

Supplementary Information The online version contains supplementary material available at https://doi.org/10.1007/978-3-031-92501-6_7.

M. Lutwin-Kawalec (✉) · S. Nesargi · A. T. Dorsey
Department of Anesthesiology and Perioperative Medicine, Nemours Children's Hospital, Wilmington, Delaware, USA

Anesthesiology and Pediatrics, Sidney Kimmel Medical College, Thomas Jefferson University, Philadelphia, Pennsylvania, USA
e-mail: Malgorzata.Lutwinkawalec@nemours.org; Susmita.Nesargi@nemours.org; Alfred.Dorsey@nemours.org

L. W. Averill
Radiology and Pediatrics, Sidney Kimmel Medical College, Thomas Jefferson University, Philadelphia, Pennsylvania, USA

Department of Radiology, Nemours Children's Hospital, Wilmington, Delaware, USA
e-mail: Lauren.averill@nemours.org

M. C. Theroux (ed.), *Anesthesia Care of Patients with Skeletal Dysplasia*,
https://doi.org/10.1007/978-3-031-92501-6_7

autosomal dominant with each adult having a 50% chance of passing the mutated gene on to each child [9]. The penetration is variable with different patients having different severities of presentation. Intelligence and life expectancy are usually average [10, 11].

7.2 Phenotype

Kniest dysplasia belongs to a group of skeletal dysplasias called type II collagenopathies. This group also includes achondrogenesis II, hypochondrogenesis, spondyloepiphyseal dysplasia congenita, Stickler arthro-ophthalmopathy, and mild dominant spondyloarthropathy [6, 9]. This group of genetic disorders shares some clinical and radiological manifestations and should be approached as a spectrum of clinical presentations of different severities. Collagen II is the predominant structural cartilage molecule in joints, bones, airway, and eyes. There is a relationship between the proportion of mutated type II collagen to normal collagen and the severity of a phenotype, with both the quantity and the quality of abnormal protein playing a role [12]. All tissues containing cartilage may be affected with respect to growth, structure, and function (Table 7.1).

Table 7.1 Conditions associated with KD [8, 12–14]

Body system involved	Condition
Head and neck	Macrocephaly
	Round and flat face with wide-set eyes
	Wide, prominent forehead
	Midface hypoplasia
	Cleft palate
	Early-onset myopia (nearsightedness)
	Retinal detachment
	Eye globe fragility
	Cataracts
	Glaucoma
	Blindness
	Recurrent middle ear infections
	Hearing loss
	Pierre Robin sequence
	Obstructive sleep apnea
	Tracheobronchomalacia
Trunk, chest, and spine	Short trunk
	Constricted chest
	Cervical spine instability (C1–2, C2–3)
	Early-onset and progressive kyphoscoliosis
Arms and legs	At birth, short limbs compared with torso (proportions change with patient's growth and progression of kyphosis)
	Rhizomelia (shortening of the upper arms and thighs in relation to the middle and distal segments of the limbs)
	Progressive large joint deformity and stiffness leading to gross motor delays
	Hand joint arthropathy leading to fine motor problems
	Hip dysplasia/dislocation
	Clubfoot
	Contractures
Gastrointestinal	Umbilical and inguinal hernias

7.3 Diagnosis and Treatment

Clinical and radiographic evaluations are usually sufficient to establish a KD diagnosis in a newborn [8, 9, 13]. Prenatal diagnosis is possible with amniocentesis and end-of-second-trimester sonography [8, 11]. Newborns with KD typically present with macrocephaly, disproportionate short stature, enlarged knee joints, and cleft palate. As mentioned in Table 7.1, not all features are present at birth, which may delay diagnosis. Salik et al. state that KD should be considered in a dysmorphic newborn with skeletal abnormalities and a cleft palate [7].

Diagnostic studies include radiographs, computed tomography (CT), and magnetic resonance imaging (MRI). Table 7.2 lists radiologic features of KD. Characteristic cloud-like calcification of the epiphyses and metaphyses of the long bones and vertebral bodies, delayed ossification of the capital femoral epiphyses, and dumbbell-shaped femurs help distinguish KD from other type II collagenopathies [15, 16]. DNA testing (pre- or postnatal) is used to confirm the mutation. Bone biopsy may be necessary in more ambiguous cases.

Care requires a multidisciplinary team approach as several organ systems are typically involved. All patients need regular follow-up and physical therapy to evaluate disease progression, preserve range of motion, and prevent progression of stiffness and permanent contractures. Some patients require surgical treatment involving stabilization of the cervical spine, scoliosis correction, surgery to prevent or release contractures, cleft palate repair, ear tube placement, or repair of retinal detachment and hernias [13].

Table 7.2 Radiologic features of KD [15, 17–21]

Bones affected	Features
Long bones	**Dumbbell shape** of long bones with broad metaphyses (Fig. 7.2)
	Irregular enlarged punctate epiphyses
	Epiphyses ossify late and may not be seen on radiographs of young patients (radiographic absence of the capital femoral epiphysis)
	Fluffy and irregular growth plates
	Loss of normal trabecular bone pattern
	Narrowing of the joint spaces
	Bilateral radial head dislocation
Spine	**Platyspondyly**: Flat vertebrae with anterior wedging of vertebral bodies (Fig. 7.1)
	Coronal clefts of thoracolumbar vertebral bodies in infants
	Kyphoscoliosis
	Bony fusion between the anterior arch of the atlas and the odontoid
Pelvis	**Trefoil-shaped pelvis**: Low broad ilia
	Wide symphysis pubis
	Narrow pelvis (Fig. 7.2)
Hip	Coxa vara: Femoral neck-shaft angle less than 120°
	Hip dislocation
	Delayed ossification of the capital femoral epiphysis
Hands and feet	Flattening and squaring-off of epiphyses of metacarpals and metatarsals (older patients)
	Disorganized appearance of the carpus
	Clubfoot

7.4 Histologic and Radiologic Findings

The histologic features of KD correlate with abnormal cartilage growth and include disorganized growth plates, unorganized-looking cartilage with a "Swiss-cheese appearance," and a diastase-resistant intracytoplasmic inclusion in chondrocytes [17, 18]. Electron microscopy images show fragmentation, deficiency, and disintegration of collagen fibrils causing large, open cyst-like spaces similar to holes in Swiss cheese [17]. Macroscopically, the cartilage feels soft. In most patients, the diagnosis of KD is established early in life based on the typical radiologic findings listed in Table 7.2. Figure 7.1 shows the typical bony findings of KD.

The dumbbell shape of long bones (Fig. 7.2) is the most characteristic finding and distinguishes KD from other connective tissue dysplasias [7]. Tubular bones assume the dumbbell morphology as their growth progresses with broad metaphyses and deformed large epiphyses. Maldjian et al. [18] report the "absence" of the

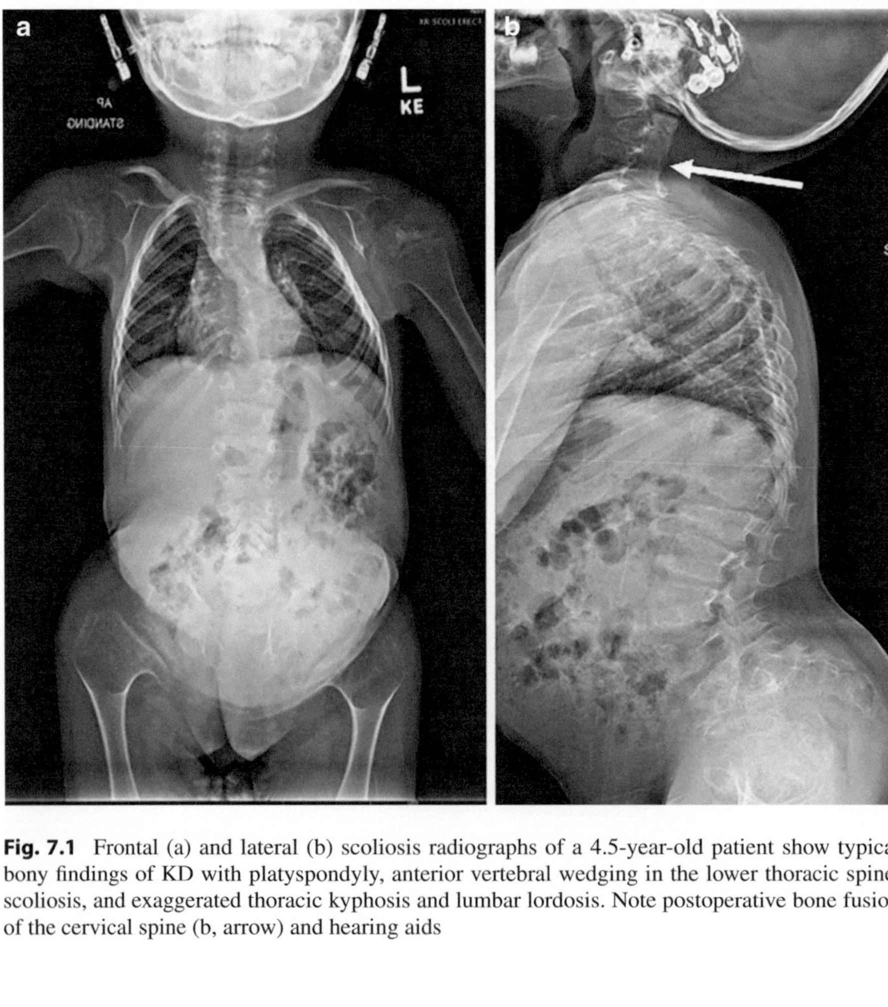

Fig. 7.1 Frontal (a) and lateral (b) scoliosis radiographs of a 4.5-year-old patient show typical bony findings of KD with platyspondyly, anterior vertebral wedging in the lower thoracic spine, scoliosis, and exaggerated thoracic kyphosis and lumbar lordosis. Note postoperative bone fusion of the cervical spine (b, arrow) and hearing aids

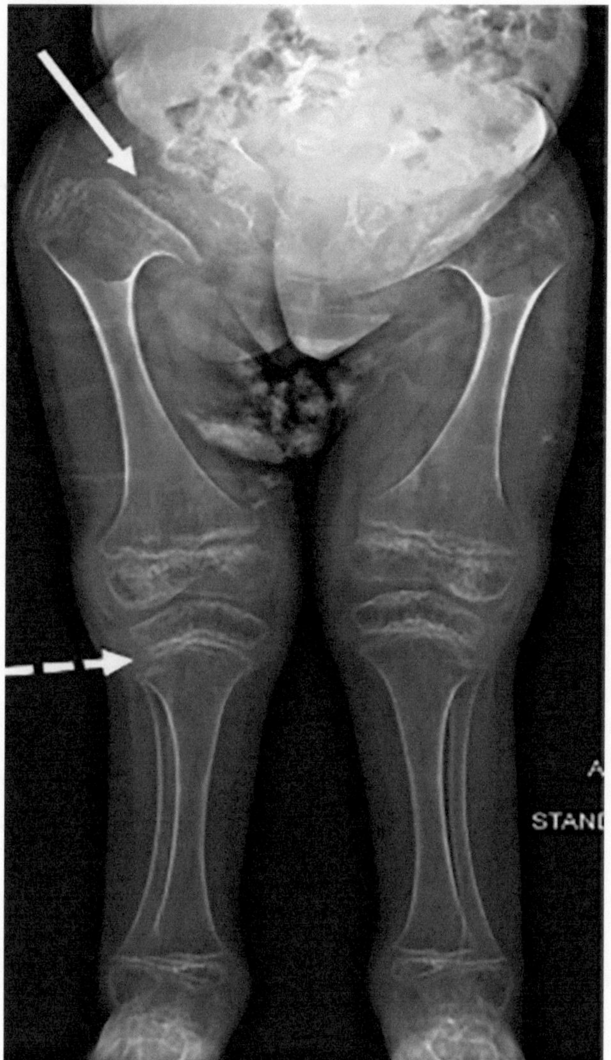

Fig. 7.2 Lower extremity radiograph of a 4.5-year-old patient shows a narrow trefoil pelvis, shortening of the long bones, dumbbell-shaped femurs, coxa vara, and genu valgum. Femoral heads are flattened and poorly ossified with a cloud-like appearance (arrow). Proximal tibia growth plates have an arched or inverted V configuration (dashed arrow). Overall, bony mineralization is reduced

capital femoral epiphysis, broadening of the metaphyses, widening of the symphysis pubis, and narrow pelvis in a newborn with KD. While the capital femoral epiphyses are not visible on radiographs, they *are* present and enlarged but ossify late. Radiographic signs of early degenerative arthritis may be present in multiple joints. Maldjian et al. describe bilateral clubfeet and bilateral radial head dislocation as common skeletal findings in these patients [18, 20].

7.5 Conditions Associated with KD

7.5.1 Trunk, Chest, and Spine

Dwarfism, short trunk, and early-onset kyphoscoliosis are caused by abnormal bone growth related to abnormal type II collagen. Cervical spine instability may cause spinal cord compression, myelopathy, and paralysis. With time, patients may develop spontaneous cervical vertebra fusion and limitation of neck movement. Signs of upper cervical myelopathy include lethargy, hypotonia, decreased muscle strength, and long tract signs such as asymmetric strength, asymmetric or increased deep tendon reflexes, and sensory changes. In patients with cervical instability or cord compression, direct laryngoscopy is best avoided to minimize cervical spine movements and thus prevent spinal cord injury. Forced cervical flexion is more injurious than extension, and this should be verified from flexion/extension imaging studies typically obtained during the preoperative appointment with the surgical team [22].

7.5.2 Extremity Abnormalities

Worsening stiffness and enlargement of various joints is common. Hand involvement usually begins in childhood and causes problems with fine motor skills requiring occupational therapy. In addition to hands, premature arthritis in shoulders, hips, and knees occurs. Coxa vara is associated with narrowing of hip joint space and limitation of movement. Flexion contractures are treated early with varus and extension osteotomies. Narrow pelvis and enlarged femoral heads predispose patients with KD to hip dysplasia, which may require pelvic osteotomies and joint replacement in certain patients. Bilateral clubfoot is more prevalent in this cohort [22].

7.6 Comorbid Conditions

7.6.1 Head and Neck

Close to 50% of patients with KD have cleft palate, which often leads to recurrent middle ear infections and delayed onset of speech. Hearing loss may occur as a result of frequent otitis media and sensorineural abnormalities. Recurrent otitis media should be treated aggressively with myringotomy and tube placement to avoid permanent hearing problems [22].

High myopia (nearsightedness) is common with significant risk of retinal detachment and blindness. Any sudden change in vision may indicate retinal detachment, which is a surgical emergency. Cataracts and glaucoma have also been reported in this patient population [22–24].

Midface hypoplasia, micrognathia, laryngo-tracheo-bronchomalacia, and limited respiratory reserve from chest constriction are common. These findings, as well as cervical spine instability, may present anesthesia providers with challenging mask ventilation, difficult intubation, hypoxia, and risk of spinal cord injury [25]. Smaller airway caliber may require smaller than calculated-for-age endotracheal tubes (ETT). In a typical patient with KD, this does not pose a problem because of their weight being well below a similarly aged average stature patient (thus the actual body mass being ventilated is significantly less, and a smaller-caliber ETT is sufficient). Collagenopathy of KD may also lead to post-extubation croup. We will cover anesthesia care in detail in the "Anesthesia and Postoperative Care" section of this chapter. Sleep apnea is common, and polysomnography is usually done in infancy. Noninvasive airway support is helpful in patients with obstructive sleep apnea, and tracheostomy may be required in some patients.

7.6.2 Malignancies

Chondrosarcomas are rare cartilage-forming neoplasms. Somatic *COL2A1* mutations characteristic for KD have also been found in chondrosarcomas. Hochart et al. reported a case of a child with KD and grade 1 sphenoethmoidal chondrosarcoma [4].

7.7 Anesthesia and Postoperative Care

Patients with KD frequently require anesthesia care for surgical procedures and imaging studies. Factors contributing to an increased perioperative risk in patients with skeletal dysplasia are summarized in Table 7.3.

Preoperative anesthesia planning starts with a review of previous anesthesia records and intubation techniques and evaluation of cervical spine stability, airway anatomy, and neurologic deficits. These findings should be documented in the chart. Increased anesthesia risk is primarily associated with difficult tracheal intubation, cervical spine instability, and risk of spinal cord injury. If the cervical spine is unstable or uncertain, intubation should be completed with utmost care to avoid injury to the spinal cord. Such intubations could be achieved using different techniques and intubation tools, and the anesthesia provider should employ the technique most likely to be successful in his/her hands. Such techniques include (1) use of flexible bronchoscope to avoid flexion/extension movements of the cervical spine, (2) use of video laryngoscope, which has been shown to minimize movement at the craniocervical junction [27], (3) combined techniques using video laryngoscope and a flexible bronchoscope as reported in the literature [28], and (4) use of a laryngeal mask airway as a conduit to introduce the flexible bronchoscope [29]. Regardless of the technique used, some precautions should be adhered to, which include use of a small shoulder roll to align the external auditory meatus with the clavicle and having a provider, preferably an orthopedic surgeon, maintain the head and neck in the neutral position. Where demonstrable instability is present, endotracheal intubation may be monitored using evoked potentials, as neuromonitoring is the most reliable

Table 7.3 Factors contributing to an increased perioperative risk in patients with skeletal dysplasia [26]

Factor	Example
Upper airway anatomy	Pierre Robin sequence
	Obstructive sleep apnea
Tracheobronchial morphology and function abnormalities	Laryngo-tracheo-bronchomalacia
	Small tracheal diameter requiring smaller ETT
	Connective tissue vulnerability predisposing to post-extubation croup
Abnormal upper cervical spine	Atlantoaxial and occipitoatlantal instability
	Previous surgical cervical fusion or auto fusion
	Difficulty with direct laryngoscopy
	Risk of paralysis with neck manipulation
Chest wall deformity	Restrictive lung disease
Kyphoscoliosis	Decreased respiratory reserve
	Risk of need for postoperative ventilation
Extremity contractures	Difficult positioning
Body habitus	Risk of peripheral nerve injuries
	Difficult intravenous cannulation
Abundant subcutaneous fat tissue	Difficult intravenous cannulation

method of assuring cervical spine integrity during intubation attempts. This is of greater significance when intubation is expected to be difficult.

Rapid development of laryngeal edema may occur in association with tracheal intubation as described by Godai et al. in a 6-year-old girl with KD undergoing gastrocnemius myotomy with resultant unplanned intensive care admission and 21 days of intubation [30]. When this patient returned for talipes equinovarus correction at age 13, intubation attempts were unsuccessful, and bronchoscopy showed false vocal cord swelling, laryngeal edema, and glottic narrowing after only two intubation attempts [31]. This patient required treatment with steroids and recovered uneventfully. Nasal endoscopy performed 1 month later showed a small larynx and narrow glottis. Cervical CT did not detect any airway narrowing from the glottis to the carina. The authors concluded that even though it was not clear if the rapid development of airway swelling was related to the patient's collagenopathy or the trauma caused by intubation attempts, perioperative rapid development of airway edema should be expected regardless of age in patients with KD. We have not experienced such perioperative edema formation ourselves, but, given the gravity and implications of such an occurrence, it is certainly worth keeping in mind. If tracheal intubation could be avoided by using a laryngeal mask airway or simply maintaining a natural airway, this may help avoid airway edema related to intubation. Preoperative airway assessment with laryngoscopy/tracheoscopy and CT is recommended, especially in young patients with KD. It is also our experience that preoperative imaging studies, especially a CT/angiogram, may help detect preexisting congenital or acquired airway abnormalities, evaluate the progression of known airway conditions, and determine the safest plan for securing the airway. For some patients with skeletal dysplasia and airway stenosis, planned non-airway surgery may need to be delayed to allow the trachea to grow, and airway reconstruction surgery prior to other surgical interventions may be needed to avoid a potentially life-threatening airway complication. To illustrate how devastating an airway complication can be, Salik et al. describe an 18-month-old boy with KD and undiagnosed congenital subglottic stenosis resulting in difficulty with tracheal intubation, cardiac arrest, and irreparable neurological damage [7]. Growth-related anatomical changes of the larynx may contribute to intubation difficulty as described by Segawa et al. [25]. An 11-year-old girl with KD with no reported anesthesia difficulties at age 6 years required multiple intubation attempts because of inability to visualize the vocal cords during direct laryngoscopy. The authors were not sure if the difficult laryngoscopy resulted from previous airway trauma or impaired airway growth as part of the collagenopathy itself [25].

Major joint contractures may result in difficulty with positioning and intravenous access. Careful positioning and padding of all pressure points are important to decrease the risk of pressure ulcers and nerve palsies.

7.8 Wound Healing

Type II collagen is known to play an important role in the early stages of wound healing and makes up to 30% of collagen in an early healing wound [32]. A reassuring case report by Husain et al. [32] describes a patient with KD who underwent multiple head and neck surgeries, namely cochlear implantation, mandibular distraction, palatoplasty, and laryngotracheal reconstruction, without any associated complications with respect to wound healing. The authors concluded that "surgery in this population can take place as indicated and should not be contraindicated or delayed." Another case report by Cuellar et al. [33] reports a 21-year-old female who underwent a reduction mammoplasty and healed without complications. The authors concluded that wound healing appears normal in this population, and it is reasonable to consider performing elective surgeries in patients with KD [33]. Our experience concurs with these authors in that we provide surgeries as indicated without fear of wound healing.

7.9 Case Illustration

A 3.5-year-old girl (weight 10.1 kg, height 73 cm) with KD, Pierre Robin sequence, cervical instability, and kyphosis presented for a posterior cervical decompression and fusion with an iliac bone graft and cervical traction application. Her medical history was as follows. Skeletal dysplasia was suspected prenatally (based on short bones seen on fetal ultrasound) and was confirmed molecularly after birth. As an infant, she developed delayed gross motor skills but was cognitively intact. At age 6 months, she underwent a cleft palate repair and a gastric tube placement. These surgeries were done at a different facility with no reported difficulties with intubation. Radiographs of the spine at age 9 months showed cervical and thoracolumbar

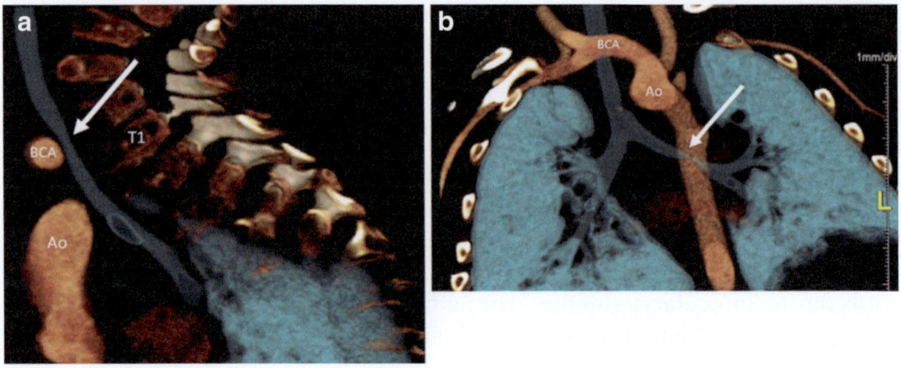

Fig. 7.3 CTA of the chest performed at age 31 months for evaluation of airway and vascular anatomy, with sagittal (**a**) and coronal (**b**) oblique three-dimensional reconstructions. Trachea and lungs are shown in teal, vessels in orange, and bone in white to orange (*BCA* brachiocephalic artery, *Ao* aortic arch, *T1* first thoracic vertebra). There is diffuse narrow caliber of the trachea with indentation by the BCA as it passes through the thoracic inlet, measuring 3 mm in anteroposterior diameter (**a**, arrow). There is also marked narrowing of the left mainstem bronchus as it passes over the descending thoracic aorta (**b**, arrow)

kyphoscoliosis. At age 18 months, she developed hearing and speech impairment, which were treated with tympanostomy tube placement and speech therapy.

At age 27 months, cervical flexion/extension imaging studies showed new atlantoaxial instability, and neurosurgery was consulted. While providing anesthesia to obtain a preoperative MRI, difficulty in advancing the ETT was encountered during an intubation attempt. The ETT was downsized to a 3.0-mm uncuffed tube from a 3.0-mm cuffed tube, resulting in successful intubation. Even though there was a leak around the ETT, the patient developed post-extubation stridor requiring an unplanned hospital admission and a 5-day course of steroids. Post-intubation croup, as mentioned before, is a known complication in patients with KD whose airway has been manipulated and is thought to be due to connective tissue vulnerability related to the *COL2A1* mutation [31].

At age 31 months, CT angiography (CTA) was performed for evaluation of airway and vascular anatomy; it showed diffusely narrow trachea with additional focal narrowing of the left mainstem bronchus as it passes over the aorta (Fig. 7.3).

Virtual bronchoscopy images generated from the same CTA exam showed deformation of the trachea lumen shape (Fig. 7.4). A flexible bronchoscopy exam done at the same time confirmed a long-segment tracheal stenosis (Fig. 7.5, Video 7.1). A rigid bronchoscopy was considered but not performed as it would require neck hyperextension and pose a significant risk of spinal cord injury.

At a follow-up visit at age 3 years, an MRI of her cervical spine was performed with sedation and natural airway. Unfortunately, the study showed worsening cervical and thoracic kyphosis with spinal cord compression. The surgical team, comprising our skeletal dysplasia orthopedic specialist and a neurosurgeon, recommended an anterior corpectomy and posterior cervical fusion with several months of

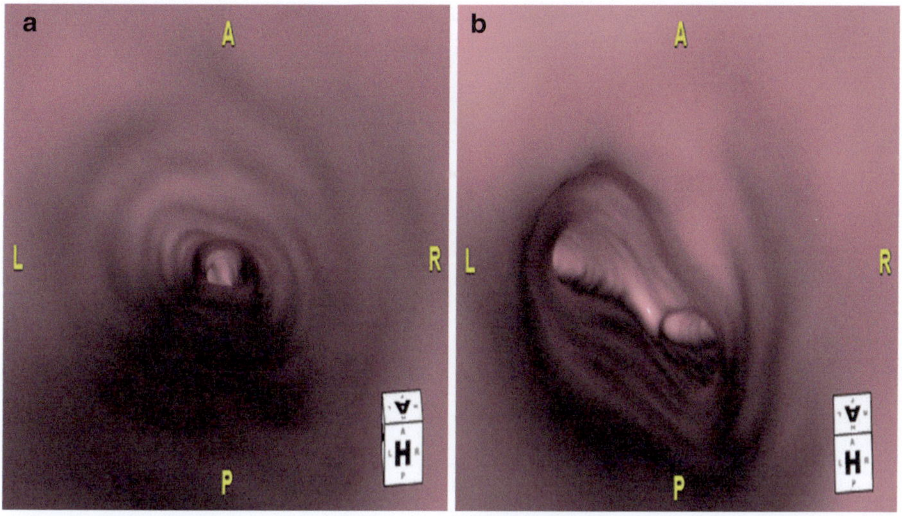

Fig. 7.4 Virtual bronchoscopy images generated from the same CTA exam just above the thoracic inlet (**a**) and above the carina (**b**) show deformation of the trachea lumen shape

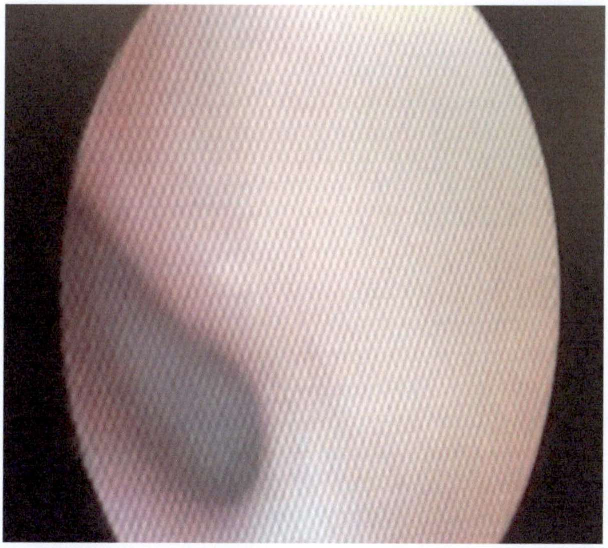

Fig. 7.5 Flexible bronchoscopy image showing marked tracheal narrowing just proximal to the carina (also see Video 7.1)

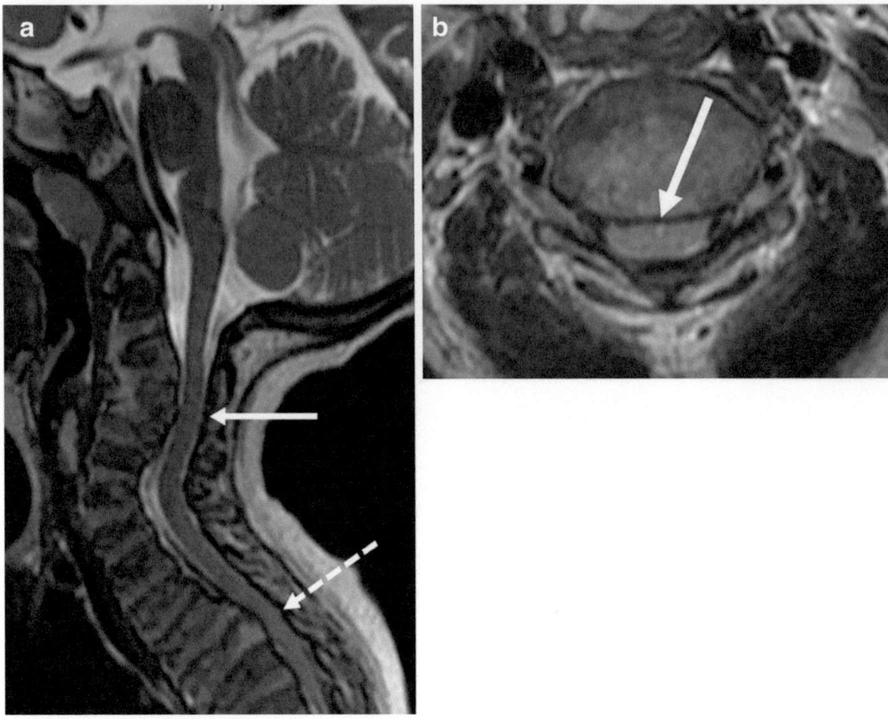

Fig. 7.6 At age 3.5 years, preoperative MRI of the cervical spine shows progressive spinal cord compression. Sagittal T2-weighted image (**a**) demonstrates features of KD, including platyspondyly, endplate irregularity, and loss of normal disc signal. There is focal kyphosis in the mid cervical spine (arrow) and upper thoracic spine (dashed arrow). Axial T2-weighted image (**b**) at the C4 level shows complete effacement of cerebrospinal fluid and flattening of the spinal cord (arrow)

postoperative cervical traction. The anesthesiologist, a member of the skeletal dysplasia team with several years of experience in taking care of patients with skeletal dysplasia, was very concerned about the significant tracheal narrowing seen on imaging and potential for a life-threatening airway complication during such an extensive surgery as anterior corpectomy combined with posterior spine fusion. There are published case reports of patients with KD and subglottic stenosis resulting in failed airway, cardiac arrest, and irreparable neurological damage [7].

As the patient remained clinically stable without any obvious deterioration in muscle strength, she was fitted in a custom-mold cervical thoracic brace for additional stability and was sent home with the plan to reevaluate in 6 months.

At her next visit (age 3.5), a spine MRI was again obtained under sedation (Fig. 7.6).

Repeat flexible bronchoscopy done at the same time showed an overall improved airway caliber with a patent but angulated subglottis and adequate tracheal diameter. Airway sizing was done with a 3.5-mm cuffed ETT placed over the scope and easily advanced into the trachea. Given the worsening cervical cord

Loss of Evoked Potentials

Management of signal changes during spine surgery

18

- Notify all members of health care team. Call a "time out"
- Loss of evoked potentials (EP) requires definitive steps to re-establish perfusion/remove mechanical cause; MEP loss for > 40 min may increase possibility of long term injury
  - Assure the presence of attending surgeon, attending anesthesiologist, senior neurologist or neurophysiologist, and experienced nurse
  - Each service: review situation, report on management and corrective actions taken
    - Surgeon: rule out mechanical causes for loss/change including traction weights
    - EP technologist: rule out technical causes for loss/change
    - Anesthesiologist: assure no neuromuscular blockade is present; reverse NMB if necessary
- Check patient positioning (neck, upper and lower extremities)
- Review the anesthetic and consider improving spinal cord perfusion by modifying:
  - Mean arterial pressure: MAP > 65 mmHg using ePHEDrine 0.1 mg/kg IV (MAX 10 mg/dose) and/or phenylephrine 0.3-1 MICROgrams/kg IV (MAX 100 MICROgrams/dose), with repeated doses as needed
  - Hemoglobin: if anemic, transfuse RBC to improve oxygen delivery
  - pH and $PaCO_2$: ensure normocarbia or slight hypercarbia ($\uparrow$ I/E ratio, $\downarrow$ PEEP)
  - Temperature: ensure normothermia
  - Check for "unintended" drugs given (e.g. neuromuscular blocker)
  - Decrease depth of anesthetic and ensure N_2O is under 50%
- Discuss feasibility of a useful wake-up test:
  - Patient is appropriate candidate if capable of following verbal commands
- Consider high-dose steroid if no improvement:
  - MethylPREDNISolone 30 mg/kg IV over one hour, then 5.4 mg/kg/hour IV for 23 hours

Loss of Evoked Potentials

Revision Feb 2020

Fig. 7.7 Society for Pediatric Anesthesia Critical Events Checklist guideline for loss of evoked potentials during spine surgery [34], used with permission

compression and the improvement in the airway diameter, after consulting all involved specialties, the decision was made to proceed with cervical decompression and fusion. The patient underwent an uneventful inhalational induction and intravenous placement. The neck was maintained in a neutral position during induction and intubation. Grade one view of the glottis was obtained with Glidescope Lo Pro (Verathon; Bothell, WA) blade size 2, but a 3.5-mm cuffed ETT was difficult to advance. A smaller 3.0-mm cuffed ETT was successfully placed. Flexible bronchoscopy performed right after intubation by the otolaryngologist showed noted tracheomalacia with thick secretions in the left mainstem bronchus. The ETT was advanced over the bronchoscope to the optimal position, and satisfactory ventilation was confirmed. Total intravenous anesthesia with propofol, ketamine, and remifentanil was used for maintenance to facilitate neuromonitoring of evoked potentials. Initial evoked potentials were obtained and noted to be abnormal likely because of preexisting cord compression. Anticipating further loss of evoked potentials during surgery, methylprednisolone 30 mg/kg was given preemptively (Society for Pediatric Anesthesia Critical Events Checklist tool for loss of evoked potentials [Fig. 7.7]). The patient was placed in the halo and then turned prone. The intraoperative course was complicated by multiple episodes of motor evoked potentials losses lasting 20–80 min in both the upper

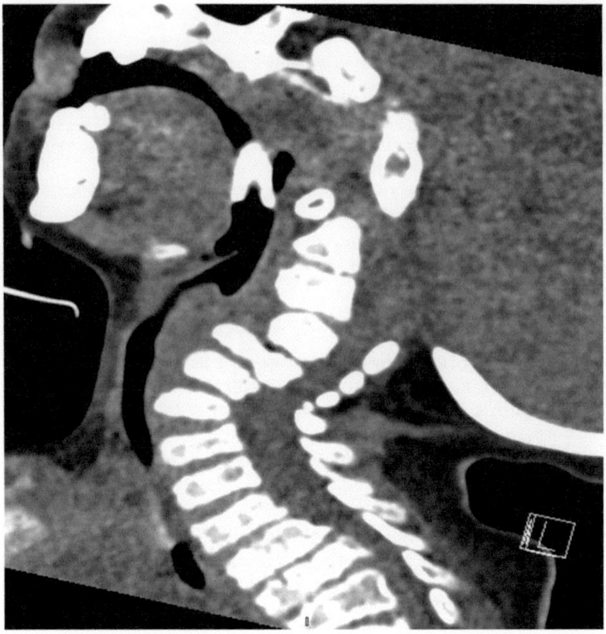

Fig. 7.8 CT image of the cervical spine shows how the trachea curves with severe lordosis. This explains the need for taping the ETT "deeper" (at 15 cm) instead of the expected 12 cm for her age

and lower extremities. During each episode, the Critical Event Checklist was used to approach signal changes (see Chap.18).

The initial signal change occurred during the C3 laminectomy/sublaminar wire placement and lasted for about 20 min. Phenylephrine infusion was started and titrated up to maintain mean arterial pressure over 80 mmHg. All traction was taken off. The second signal change occurred during an attempted sublaminar wire placement under C2 and lasted until posterior decompression of C2 was performed. After the second episode, an infusion of methylprednisolone was added at 5.4 mg/kg/h and continued for 23 h. With these interventions, evoked potentials returned to baseline by the end of surgery. The procedure required 7.5 h to complete. Given the patient's personal history of post-extubation stridor and need for an MRI of the cervical spine on the following day, the decision was made to keep the patient intubated, and she was transferred to the pediatric intensive care unit (PICU) for postoperative monitoring and frequent neurologic evaluations.

The PICU course was complicated by unintentional extubation after the ETT was determined to be "too deep" based on the patient's age and was pulled back. This required an emergent call to anesthesia and bedside ETT replacement using flexible bronchoscopy. It is of utmost importance to keep in mind the distorted airway anatomy in patients with KD (Fig. 7.8) and exercise caution when "adjusting" tube position. We recommend using flexible bronchoscopy for this purpose as it allows real-time determination of ETT tip position in relation to the carina.

The patient was successfully extubated in the operating room the following day after completion of her previously scheduled MRI of the cervical spine. Cervical traction was continued for 3 months. After that, the patient was placed in a hard cervical collar for an additional 2 months.

Eight months after surgery, the patient was doing well and walking significantly better. Fine motor abilities in her hands improved to the point that she no longer requires occupational therapy.

This case highlights the airway and anesthetic challenges encountered in the care of patients with KD and an ever-changing clinical situation that requires adjustments in surgical and anesthesia planning and an interdisciplinary approach. Close and ongoing communication between the different specialties involved is critical. This case also illustrates why this patient population is best cared for at tertiary-care hospitals.

References

1. Winterpacht A, Superti-Furga A, Schwarze U, Stoss H, Steinmann B, Spranger J, et al. The deletion of six amino acids at the C-terminus of the alpha 1 (II) chain causes overmodification of type II and type XI collagen: further evidence for the association between small deletions in COL2A1 and Kniest dysplasia. J Med Genet. 1996;33:649–54.
2. Cole WG. Abnormal skeletal growth in Kniest dysplasia caused by type II collagen mutations. Clin Orthop Relat Res. 1997;341:162–9.
3. Wilkin DJ, Artz AS, South S, Lachman RS, Rimoin DL, Wilcox WR, et al. Small deletions in the type II collagen triple helix produce Kniest dysplasia. Am J Med Genet. 1999;85:105–12.
4. Hochart A, Dieux A, Coucke P, Fron D, Fayoux P, Labalette P, et al. Association between Kniest dysplasia and chondrosarcoma in a child. Am J Med Genet A. 2015;167A:3204–8.
5. Spranger J, Winterpacht A, Zabel B. Kniest dysplasia: Dr. W. Kniest, his patient, the molecular defect. Am J Med Genet. 1997;69:79–84.
6. Weis MA, Wilkin DJ, Kim HJ, Wilcox WR, Lachman RS, Rimoin DL, et al. Structurally abnormal type II collagen in a severe form of Kniest dysplasia caused by an exon 24 skipping mutation. J Biol Chem. 1998;273:4761–8.
7. Salik I, Kubal K, Barst S. Anesthetic implications of a patient with Kniest dysplasia and mitochondrial disease: a case report. A A Pract. 2019;12:267–9.
8. The Nemours Foundation [Internet]. Kniest dysplasia. 2022. https://www.nemours.org/services/kniest-dysplasia/about-kniest-dysplasia.html.
9. Spranger J, Winterpacht A, Zabel B. The type II collagenopathies: a spectrum of chondrodysplasias. Eur J Pediatr. 1994;153:56–65.
10. Benumof J. Anesthesia & uncommon diseases. 4th ed. Philadelphia: W.B Saunders; 1998.
11. The Children's Hospital of Philadelphia [Internet]. Kniest dysplasia. 2022. https://www.chop.edu/conditions-diseases/kniest-dysplasia.
12. Hicks J, De Jong A, Barrish J, Zhu SH, Popek E. Tracheomalacia in a neonate with Kniest dysplasia: histopathologic and ultrastructural features. Ultrastruct Pathol. 2001;25:79–83.
13. National Organization for Rare Disorders [Internet]. Kniest dysplasia. 2021. https://rarediseases.org/gard-rare-disease/kniest-dysplasia/.
14. Jhamb T, Masood H, Arigo J, Rossouw PE. Orthodontic treatment in a patient with Kniest dysplasia: a case study and review of literature. Cleft Palate Craniofac J. 2019;56:1393–403.
15. Taybi H. Radiology of syndromes, metabolic disorders, and skeletal dysplasias. 4th ed. St. Louis: Mosby; 1996.

16. Friede H, Matalon R, Harris V, Rosenthal IM. Craniofacial and mucopolysaccharide abnormalities in Kniest dysplasia. J Craniofac Genet Dev Biol. 1985;5:267–76.
17. Gilbert-Barnes E, Langer LO, Opitz JM Jr, Laxova R, Sotelo-Arila C. Kniest dysplasia: radiologic, histopathological, and scanning electronmicroscopic findings. Am J Med Genet. 1996;63:34–45.
18. Maldjian C, Chew FS, Klein R, Bonakdarpour A, McCarthy J, Kelly J. Kniest dysplasia: new radiographic features in the skeleton. Radiol Case Rep. 2015;2:72–7.
19. Lachman RS, Rimoin DL, Hollister DW, Dorst JP, Siggers DC, McAlister W, et al. The Kniest syndrome. Am J Roentgenol Radium Therapy, Nucl Med. 1975;123:805–14.
20. Oestreich AE, Prenger EC. MR demonstrates cartilaginous megaepiphyses of the hips in Kniest dysplasia of the young child. Pediatr Radiol. 1992;22:302–3.
21. Brill PW, Kim HJ, Beratis NG, Hirschhorn K. Skeletal abnormalities in the Kniest syndrome with mucopolysacchariduria. Am J Roentgenol Radium Therapy Nucl Med. 1975;125:731–8.
22. Little People of America [Internet]. Kniest dysplasia natural history. 2009. https://www.lpaonline.org/assets/documents/NH%20Kniest%20dysplasia1.pdf.
23. Makino I, Watanabe M, Okamoto N, Ohguro N, Fukuda M, Emi K. A case of retinal detachment in Kniest dysplasia treated with vitreous surgery. Nippon Ganka Gakkai Zasshi. 1997;101:734–7.
24. Kagotani Y, Takao K, Nomura K, Okubo K. Two cases of Kniest dysplasia--ocular manifestations. Nippon Ganka Gakkai Zasshi. 1995;99:376–83.
25. Segawa Y, Enomoto H, Nakagawa T, Iijima Y. Growth causes difficult tracheal intubation in a patient with Kniest dysplasia. J Anesth. 2001;15:104–5.
26. White KK, Bompadre V, Goldberg MJ, Bober MB, Cho TJ, Hoover-Fong JE, et al. Best practices in peri-operative management of patients with skeletal dysplasias. Am J Med Genet A. 2017;173:2584–95.
27. Agrawal S, Salunke P, Gupta S, Swain A, Jangra K, Panda N, et al. Fiberoptic bronchoscopy versus video laryngoscopy guided intubation in patients with craniovertebral junction instability: a cinefluroscopic comparison. Surg Neurol Int. 2021;12:92.
28. Khan MF, Shamim F, Slote MU, Salim B, Abbas SA. Combined use of a videolaryngoscope and a flexible bronchoscope for awake tracheal intubation when front-of-neck airway is not an option. Anaesth Rep. 2021;9:12–5.
29. Wong DT, Yang JJ, Mak HY, Jagannathan N. Use of intubation introducers through a supraglottic airway to facilitate tracheal intubation: a brief review. Can J Anaesth. 2012;59:704–15.
30. Godai K, Hasegawa-Moriyama M, Masuda M, Matsunaga A, Isowaki S, Kanmura Y. Rapid development of airway edema during tracheal intubation in a patient with Kniest dysplasia. Anesth Resusc. 2012;48:53–4.
31. Hasegawa-Moriyama M, Iwasaki T, Mukaihara K, Masuda M, Kanmura Y. Unsuccessful tracheal intubation in a patient with Kniest dysplasia undergoing repeated general anesthesia: a case report. JA Clin Rep. 2018;4:41.
32. Husain Q, Cho J, Neugarten J, Modi VK. Surgery of the head and neck in patient with Kniest dysplasia: is wound healing an issue? Int J Pediatr Otorhinolaryngol. 2017;93:97–9.
33. Cuellar TA, Erdmann-Sager J. Reduction mammaplasty in a patient with Kniest dysplasia: case report and literature review. Plast Reconstr Surg Glob Open. 2020;8:e2649.
34. Society for Pediatric Anesthesia Critical Event Checklist (www.pedsanesthesia.org) [Internet]. https://pedsanesthesia.org/wp-content/uploads/2020/11/SPAPediCrisisChecklists-Nov2020.pdf.

Stickler Syndrome

8

Sheaba L. Varghese, Sharon Luo, and Nathalie Peiris

8.1 Introduction

Stickler syndrome is a genetic disorder affecting collagen formation often diagnosed in infancy. Patients present with facial abnormalities, joint problems, and visual and hearing disturbances, which has resulted in the syndrome being referred to as arthro-ophthalmopathy. The syndrome is most commonly inherited in an autosomal-dominant manner and exhibits genetic heterogeneity wherein several genes have been implicated [1]. Collagen plays a vital role in connective tissue development, so it is not surprising that these patients have significant joint involvement secondary to mutations in the formation of collagen.

8.2 Genetics/Inheritance

Stickler syndrome is a genetically complex disease that affects collagen formation. In particular, the collagens affected are type II, IX, and XI, which are vitreous, skeleton, and inner ear collagens [1]. Stickler syndrome is thought to affect 1 in every 7500–9000 newborns [2]. Due to genetic heterogeneity, there are currently six known genetic subtypes of Stickler syndrome [3]. See Table 8.1 for summary.

S. L. Varghese (✉)
Anesthesiology and Pediatrics, University of Cincinnati, Cincinnati, Ohio, USA

Cincinnati Children's Hospital Medical Center, Cincinnati, Ohio, USA
e-mail: Sheaba.varghese@cchmc.org

S. Luo · N. Peiris
Department of Anesthesiology and Perioperative Medicine, Nemours Children's Hospital, Wilmington, Delaware, USA

Anesthesiology and Pediatrics, Sidney Kimmel Medical College, Thomas Jefferson University, Philadelphia, Pennsylvania, USA
e-mail: Sharon.Luo@nemours.org; Nathalie.Peiris@nemours.org

M. C. Theroux (ed.), *Anesthesia Care of Patients with Skeletal Dysplasia*,
https://doi.org/10.1007/978-3-031-92501-6_8

Table 8.1 Summary of subtypes of Stickler syndrome [3]

Type	Gene affected	Genetic inheritance	Characteristic features
1	*COL2A1*	Autosomal dominant	– Membranous congenital vitreous anomaly, congenital megalopthalmos, deafness, arthropathy, cleft palate – Also variant with ocular symptoms only
2	*COL11A1*	Autosomal dominant/ autosomal recessive	– Beaded congenital vitreous anomaly, congenital megalopthalmos, deafness, arthropathy, cleft palate
3	*COL11A2*	Autosomal dominant	– Normal vitreous and ocular phenotype, deafness, arthropathy, cleft palate
4	*COL9A1*	Autosomal recessive	– Sensorineural deafness, myopia, vitreoretinopathy, epiphyseal dysplasia
5	*COL9A2*		
6	*COL9A3*		
Other	Unknown	Unknown	Hypoplastic vitreous, deafness, arthropathy, cleft palate

The subtypes of Stickler syndrome differ among the vitreous presentation. The syndrome is subcategorized into three different types based on their ocular phenotype and molecular linkage [4].

The most common form is Stickler syndrome type I, which accounts for about 80–90% of Stickler syndrome cases and exhibit milder symptoms [4, 5]. There are also patients with Stickler syndrome type 1 that only have ocular manifestations due to alternate splicing of the *COL2A1* gene [1].

8.3 Patient Presentation

Facial findings include prominent eyes, small nose, receding chin, and often a cleft palate. Patients may also have a flat midface with depressed nasal bridge, reduced nasal protrusion, anteverted nares, and micrognathia. These facial features become less obvious as the child ages [5]. Visual complications are the most serious manifestations with retinal detachment being the most common. Stickler syndrome is the most common inherited cause of retinal detachment in childhood [5]. Individuals are commonly myopic and at high risk for ocular complications, including vitreous abnormalities, glaucoma, and cataracts, which are congenital and nonprogressive [1]. Hearing impairment is usually sensorineural, affects high tone, and can be progressive. Skeletal findings can include scoliosis, slipped epiphysis, or even osteoarthritis presenting before age 40 years [6].

Table 8.2 Proposed diagnostic criteria for Stickler syndrome type 1 [4]

Abnormality	
Ocular	Characteristic vitreous changes or retinal abnormalities (retinal detachment, tear, lattice degeneration)
Craniofacial	Cleft palate, bifid uvula
	Characteristic facies (malar hypoplasia, broad/flat nasal bridge, micrognathia)
Auditory	High frequency sensorineural hearing loss
Skeletal	Femoral head failure (slipped epiphysis, Legg-Perthes disease)
	Radiographically demonstrated osteoarthritis before age of 40
	Scoliosis, spondylolisthesis, Scheurmann-like kyphotic deformity
Family history	1st degree relative with presence of COL2A1, COL11A or COL11A2 mutation

8.4 Phenotype

Stickler syndrome patients have variable phenotypic expression. Thus, different members of the same family may present with different manifestation and severity. A brief summary of characteristic features for each subtype can be found in Table 8.2.

8.4.1 Ocular Symptoms

Patients with Stickler syndrome are at high risk for vitreous abnormalities, myopia, and retinal detachment. Congenital cataracts and risk of glaucoma due to abnormal drainage of the anterior chamber of the eye can also be present. The risk of retinal detachment is highest in Stickler syndrome type I and II—this typically will occur between ages 10 and 30 years and is a significant cause of blindness in children [1]. Stickler syndrome type III (which was formerly known as non-ocular Stickler syndrome) is the only known subtype without ocular manifestations due to *COL11A2* not having any expression in the eye.

8.4.2 Craniofacial Symptoms

Forty percent of patients with Stickler syndrome present with cleft palate, and 24% have Pierre Robin sequence (PRS) [7] with associated micro- or retrognathia. Other common manifestations include midface hypoplasia.

8.4.3 Auditory Symptoms

Sensorineural hearing loss is the most common auditory abnormality found in patients with Stickler syndrome and typically affects higher frequencies.

8.4.4 Skeletal Symptoms and Radiographic Findings

Premature osteoarthritis is a common symptom found in Stickler syndrome. Joint hypermobility can be seen in younger children, but then resolves as they age. Legg-Calvé-Perthes disease causes remodeling of the femoral head due to vascular supply disruption and has been found to be associated with patients with Stickler syndrome type I [3]. Spinal abnormalities such as scoliosis, kyphosis, spondylolisthesis have also been seen in Stickler syndrome.

8.5 Diagnosis

Stickler syndrome should be suspected if patients present with craniofacial abnormalities, cleft palate, hearing impairment, retinal detachment, and skeletal abnormalities.

Stickler syndrome typically is diagnosed using a combination of clinical manifestations, family history, and genetic testing. Clinical diagnostic criteria have been proposed for Stickler syndrome type I but have not been validated. Points are given to clinical features, family history, and molecular data [6]. Stickler syndrome is considered if the points combine to ≥5 with one major finding and no possible alternative diagnosis [6].

See Table 8.2 for a list of proposed clinical diagnostic criteria for Stickler syndrome.

Unfortunately, these criteria may miss those that have type I ocular only or autosomal-recessive Sticker syndrome. Thus, if there is high clinical suspicion, genetic testing should be performed.

8.6 Anesthetic Considerations

The anesthetic management of patients with Stickler syndrome is challenging due to the limited exposure and experience of most anesthesiologists to these patients. Characteristic features of Stickler syndrome include congenital abnormalities of the eyes, micrognathia, cleft palate, very flexible joints, and mitral valve prolapse [1]. Since Stickler syndrome is an autosomal-dominant hereditary progressive arthro-ophthalmopathy [8], most patients are diagnosed during infancy or childhood. Sometimes phenotypic variability, different presentation of individual subtypes, and overlap with other syndromes may contribute to delayed diagnosis.

Several systems are affected by Stickler syndrome, and a thorough history and physical exam is required. During the preoperative evaluation, special evaluation of craniofacial features and cardiac exam are vital. Preoperative anesthetic assessment focuses on airway assessment, degree of neck movement, other syndromic features, and feelings of fatigue or an elevated heartbeat or presence of murmur.

Zimmermann et al. [7] recently reviewed airway management of a case series of 502 patients with known Stickler syndrome. In that retrospective study, 4% of patients with Stickler syndrome type I required multiple attempts for intubation, 2.1% percent required conversion to a different device, and 1.2% needed an advanced airway technique. Patients with Stickler syndrome type 2 and spondylo-epiphyseal dysplasia congenita (SEDC) have less incidence of needing multiple attempts for intubation or advanced airway techniques. Based on the Zimmermann et al. report and other similar studies [9, 10], a thorough airway examination is warranted. The presence of previous cleft repair and receding jaw raises high alert for a potentially challenging intubation.

Stickler syndrome also overlaps with other syndromes, like Pierre Robin [11]. When there are features suggesting other syndrome clusters, anesthetic considerations should also include ones indicated by those other syndromes, such as PRS. Although neck involvement in patients with Stickler and arthropathy is uncommon, careful examination of neck movements is performed to prevent problems during direct laryngoscopy and tracheal intubation.

Approximately, 50% of patients with Stickler syndrome may have associated congenital cardiac anomalies with the most common anomaly being mitral valve prolapse. The presence of mitral valve prolapse may make some patients have fatigue or palpitations. A thorough cardiac history and physical should be performed to allow identification of such cardiac anomalies to evaluate if the patient is symptomatic or not. If they have been diagnosed with mitral valve prolapse, medications should be reviewed and electrocardiogram obtained as these patients can present with arrhythmias. If the patient has experienced arrhythmias, then beta blockers would need to be continued in the perioperative period.

Intraoperative anesthetic management should include experienced anesthesia provider availability, especially if patients have previous cleft palate repair or receding jaw. Intubation preparation should include availability of alternative intubation devices, such as video-laryngoscope or fiberoptic scope. As in all cases of anticipated difficult airway, sufficient supplies for mask ventilation and intubation should be available. To assist with mask ventilation, ensure appropriately sized oral airways and supraglottic airways are available in case issues with ventilation arise prior to securing the airway with an endotracheal tube. Since the experience of the anesthesiologist may be limited with this patient population, there should be sufficient help available to the team to safely secure the airway.

Because a certain percentage of patients may have glaucoma [12], succinylcholine and other drugs that may increase intraocular pressure should be avoided. Both general anesthesia and regional anesthesia work equally well and there are no specific indications or contraindications for anesthetic agents.

Due to arthropathy, positioning for surgery may prove to be difficult. It may be appropriate for the patients to position themselves to prevent damage to arthritic joints. Extra precautions should be taken when positioning the patient for surgery if there is neck positioning that makes the patient uncomfortable.

In the post-anesthesia unit, a small percentage of patients may require a chin lift or jaw thrust to relieve airway obstruction; however, most patients do not experience significant issues [7].

8.7 Prognosis

The prognosis can be favorable if diagnosed early, but those who do not have the benefit of early diagnosis are at risk of devastating outcomes, including blindness. Patients will require routine hearing and vision tests to monitor visual disturbances and changes. Patients who have inherited the disease in an autosomal-dominant manner have likely inherited the disease from a parent [6]. As a result, genetic counseling should be part of the management to ensure patients are educated on the implications of the syndrome and need for long-term follow-up.

There have been studies to investigate preventing retinal detachments in these patients. One such study was the Cambridge prophylactic cryotherapy protocol, which evaluated the safety and efficacy of prophylactic treatment of retinal tears in patients with Stickler syndrome type 1 [13].

References

1. Boothe M, Morris R, Robin N. Stickler syndrome: a review of clinical manifestations and the genetics evaluation. J Pers Med. 2020;10:105.
2. Nowak CB. Genetics and hearing loss: a review of Stickler syndrome. J Commun Disord. 1998;31(437–53):453–4.
3. McArthur N, Rehm A, Shenker N, Richards AJ, McNinch AM, Poulson AV, et al. Stickler syndrome in children: a radiological review. Clin Radiol. 2018;73(678):e13–8.
4. Rose PS, Levy HP, Liberfarb RM, Davis J, Szymko-Bennett Y, Rubin BI, et al. Stickler syndrome: clinical characteristics and diagnostic criteria. Am J Med Genet A. 2005;138A:199–207.
5. Snead MP, Yates JR. Clinical and molecular genetics of Stickler syndrome. J Med Genet. 1999;36:353–9.
6. Robin NH, Moran RT, Ala-Kokko L. Stickler syndrome. 2000 Jun 9 [updated 2021 may 6]. In: Adam MP, Ardinger HH, Pagon RA, et al., editors. GeneReviews® [internet]. Seattle: University of Washington, Seattle; 1993–2021.
7. Zimmermann J, Stubbs DJ, Richards AJ, Alexander P, McNinch AM, Matta B, et al. Stickler syndrome: airway complications in a case series of 502 patients. Anesth Analg. 2021;132:202–9.
8. Stickler GB, Hughes W, Houchin P. Clinical features of hereditary progressive arthro-ophthalmopathy (Stickler syndrome): a survey. Genet Med. 2001;3:192–6.
9. Varandas J, Brito S, Rodrigues M, Cavaleiro C, Pimenta A, Machado HS. Airway management in Stickler syndrome patient: a case report. J Anesth Clin Res. 2017;8:2.
10. Küçükyavuz Z, Ozkaynak O, Tüzüner AM, Kişnişçi R. Difficulties in anesthetic management of patients with micrognathia: report of a patient with Stickler syndrome. Oral Surg Oral Med Oral Pathol Oral Radiol Endod. 2006;102:e33–6.
11. Cladis F, Kumar A, Grunwaldt L, Otteson T, Ford M, Losee JE. Pierre Robin sequence: a perioperative review. Anesth Analg. 2014;119:400–12.
12. Shenoy BH, Mandal AK. Stickler syndrome associated with congenital glaucoma. Lancet. 2013;381:422.
13. Fincham GS, Pasea L, Carroll C, McNinch AM, Poulson AV, Richards AJ, et al. Prevention of retinal detachment in Stickler syndrome: the Cambridge prophylactic cryotherapy protocol. Ophthalmology. 2014;121:1588–97.

Diastrophic Dysplasia

9

Sabina Dicindio and Abraham Oommen

9.1 Introduction

Diastrophic dysplasia (DTD) is an autosomal-recessive inherited chondrodysplasia. It has variable penetrance and a wide spectrum of clinical manifestations. The spectrum of presentation includes lethal forms, with death occurring soon after birth due to cardiorespiratory insufficiency, and mild forms with average stature. Most have a normal life expectancy, with affected individuals having short stature with unusually short arms and legs. Spinal deformities, osteoarthritis, joint contractures, club feet, and cleft palate are characteristic.

9.2 Genetics

Diastrophic dysplasia is caused by mutations in the DTD sulfate transporter gene (*DTDSD*; also known as solute carrier family 26, member 2 [*SLC26A2*]). This gene has been mapped to chromosome 5q32-q33.1 [1]. Reduced sulfate transport into chondrocytes results in decreased sulfation of proteoglycans, which causes structural compromise of cartilage and bone. Although DTD is extremely rare, Finland has a high prevalence of 1 in 22,000–30,000; elsewhere in the world, the incidence is 1 in 100,000, and in the USA, incidence is 1:500,000 [2–4]. The frequency of DTD in Finland has been decreasing since 1960. New mutations may account for 5% of cases, and the frequency is equal in males and females.

S. Dicindio (✉) · A. Oommen
Anesthesiology and Pediatrics, Sidney Kimmel Medical College,
Thomas Jefferson University, Philadelphia, Pennsylvania, USA

Department of Anesthesiology and Perioperative Medicine, Nemours Children's Hospital,
Wilmington, Delaware, USA
e-mail: sabina.dicindio@nemours.org; Abraham.oommen@nemours.org

M. C. Theroux (ed.), *Anesthesia Care of Patients with Skeletal Dysplasia*,
https://doi.org/10.1007/978-3-031-92501-6_9

9.3 Phenotype

Though DTD is a disorder with variable penetrance and a wide spectrum of clinical manifestations, classic DTD-affected individuals have a normal sized head, short stature with short limbs, small chest, and spinal deformities. Extremity involvement is rhizomelic in 40% and mesomelic in 29% of patients [2]. In addition, hand and foot deformities occur in most patients. The fingers are short, and hitchhiker's thumb is pathognomonic for DTD (Fig. 9.1). Ulnar deviation of the fingers also is noted. Ankle and foot deformities are associated with clubfeet, and a gap between the first and second toe is observed commonly. Joint deformities, contractures, and absent anterior cruciate ligament in the lower extremities may manifest as valgus deformity, joint laxity, and patellar subluxation [3].

Patients with DTD may have a cleft palate, with an incidence of 27–64%, as well as micrognathia [3, 5]. Pathognomonic for DTD is cauliflower ears, with a reported incidence of 85% [2]. This is due to swelling of the ear, which develops in infancy; with resolution of the swelling, the cartilage thickens and deforms. Eventually, the pinnae may become calcified or even ossified. Typically, older patients have rock-hard masses in the superior portions of the ear, which are easily palpable.

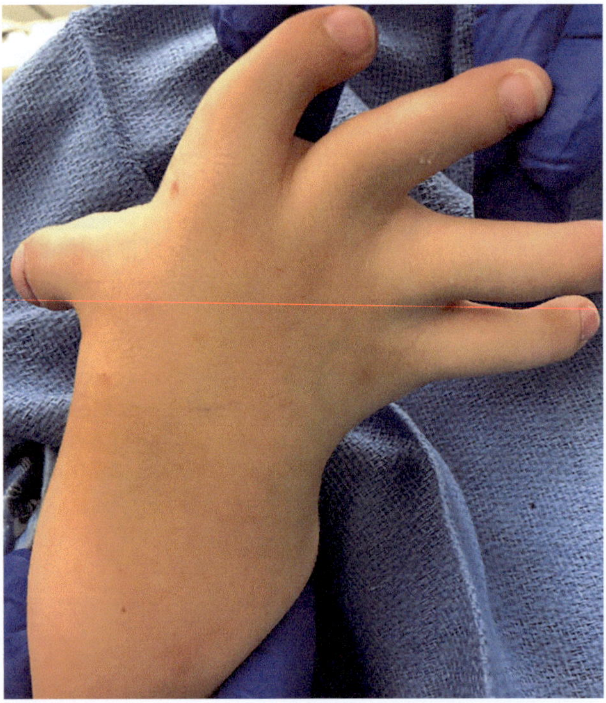

Fig. 9.1 The Hitchhiker's thumb is an abnormally positioned thumb which can bend backward when extended (resembling the gesture of a hitchhiker). It is a notable feature of diastrophic dysplasia

Spinal deformities in the DTD population are variable. In infancy, cervical kyphosis has an incidence of 25–79%; this tends to resolve with growth [3, 6, 7]. Adult patients with DTD may have cervical kyphosis, exaggerated lumbar lordosis, and/or scoliosis, 37–90% [3, 7–9].

9.4 Radiologic Findings

Radiologic findings include normal skull size with disproportionate short skeleton. Patients may have delayed epiphyseal ossification, which appears flattened and irregular on X-ray. The long bones are short, and the metaphases are widened.

The short and small first metacarpal bones cause abduction of the thumb, described as hitchhiker's thumb. The phalanges may be widely spaced and small, fused at the proximal interphalangeal joints, and deviated toward the ulna due to ulnar shortening. Similarly, foot deformities are due to the metatarsals having abnormal deviation and fusion, leading to clubfeet. Subluxation, dysplasia, and degenerative osteoarthritis of the hip are common. The sacrum may be tilted posteriorly, and the iliac bones can be hypoplastic. The knees, too, may have flexion contractures. Excessive valgus with lateral dislocation of the patella may occur, while genu varum is uncommon. Eighty percent of patients present with some degree of spinal curvature [2]. Cervical spine findings include kyphosis (24–50%), spina bifida occulta, anterior hypoplasia of C3–C5, and hyperplasia and dysmorphism of the odontoid [8]. One-half to one-third of patients have cervical kyphosis at birth. Cervical lordosis can be observed in the adult population (Fig. 9.2). Scoliosis is not present at birth, but it tends to develop within the first year of life (37–88%); it becomes progressively severe with weight-bearing [3, 7]. Lumbar lordosis is usually significant. The cervical and thoracolumbar spinal canals are narrower than in the non-SD population. Vertebral bodies are flattened and can be hypoplastic and displaced. In addition, the vertebral interpedicular distance is decreased.

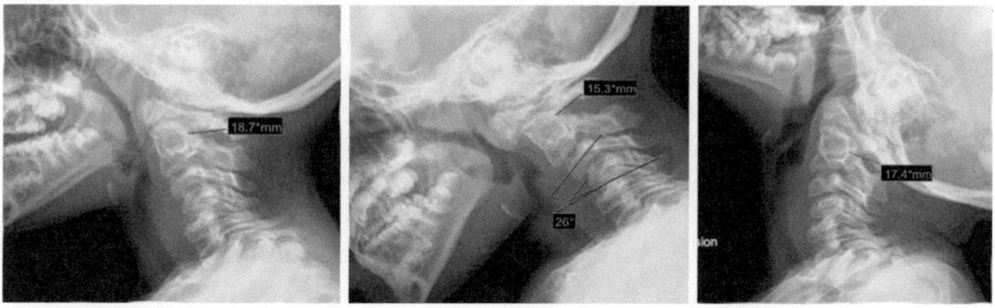

Fig. 9.2 Cervical spine X-rays: neutral, flexion, and extension. Cervical spine to the level of C6: In neutral spondylolisthesis of C3 on C4, retrolisthesis of C4 on C5 and spondylolisthesis of C6 on C7, exaggerated cervical lordosis is at apex of C6 and focal kyphosis apex at C4, the odontoid is hypoplastic, and there is 1.3-mm narrowing of the neural canal at C1–2 between flexion and extension and 3.2-mm narrowing between neutral and flexion at C1–2

9.5 Co-morbid Conditions

Most patients with DTD have normal intelligence and normal life expectancy. The associated co-morbidities are due to the abnormal growth of cartilage and bone. Though rare, lethal forms with death soon after birth due to cardiorespiratory insufficiency or spinal cord compression have been reported [3]. More commonly, patients have respiratory insufficiency due to a small rib cage. In addition, the laryngeal, tracheal, and bronchial cartilage is abnormally soft. This contributes to narrowing of respiratory passages as well as sleep disordered breathing (SDB).

The ligaments, tendons, and joint capsule are shorter and tighter, resulting in reduced joint mobility. Joint contractures worsen with age. Due to brachydactyly and symphalangism (ankylosis or synostosis of inter phalangeal joints), patients with DTD have limited and restricted movement of the finger joints. In addition, difficulty opposing the thumb results due to its proximal location. Club feet and shortened tendons cause many adults not to be able to put their heels on the ground and results in metatarsal and toe walking. Degenerative changes of the hip can lead to anterior tilting of the pelvis and exacerbate lumbar lordosis. In addition, paradoxical patellar flexion may cause difficulty with walking.

The cervical kyphosis of infancy usually resolves with extensor muscles of the cervical spine gaining strength from holding the head up [8]. However, in some patients, cervical kyphosis may progress to spinal cord compression with sudden death (5%) [2]. Vertebral body hypoplasia and displacement usually resolves with growth, and neurologic symptoms are rare. The narrowed spinal canal can predispose to neurologic compromise due to cord compression.

9.6 Clinical Presentation

At Nemours Children's Health, Delaware Valley, at the time this chapter was written, there were 44 patients with DTD who had 267 surgeries. From this cohort of patients with DTD, we are able to describe the demographics and co-morbid conditions presented in Table 9.1. The youngest patient was anesthetized at age 6 months and the oldest at age 26 years. Commonly observed co-morbidities included cervical kyphosis, kyphoscoliosis, reactive airway disease (RAD), laryngo-tracheo-broncho-malacia, SDB, and clubfeet (Fig. 9.3).

The most common surgery was orthopedic (83.5%). Other procedures in which anesthesia was needed included medical imaging (11.6%), otolaryngology (2.9%), general surgery (1.1%), urology (0.3%), and neurosurgery (0.3%) (Fig. 9.4). The most common orthopedic procedure was for serial cast changes and spine fusion of the thoracolumbar spine (n = 151) followed by an orthopedic extremity surgery (n = 70). Eight of the 44 patients needed cervical spine fusion due to instability.

Table 9.1 Mean and standard deviation for age, height, and weight

Male n = 26 (%)		Female n = 18 (%)
Age (years)	6.7 (4.2)	6.3 (4.3)
Weight (kg)	18.1 (19.7)	17.5 (21.5)
Height (cm)	90.9 (21.1)	89.1 (21.9)

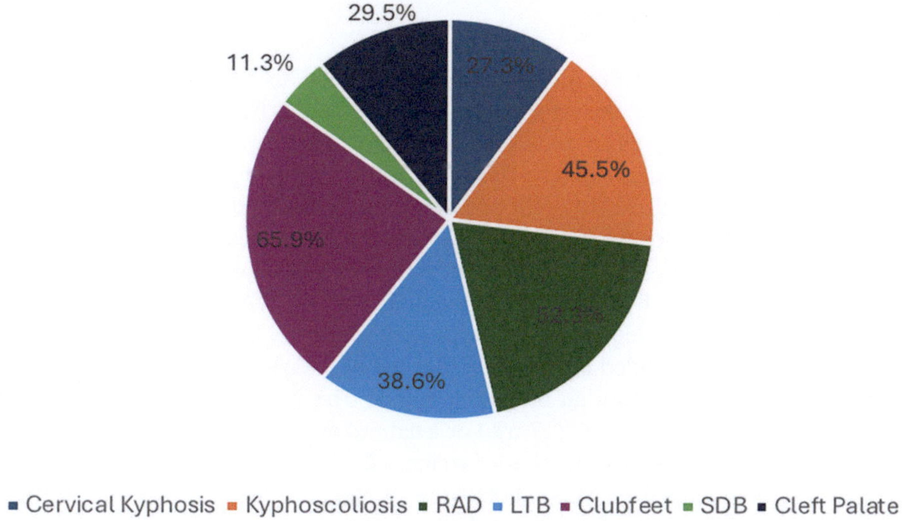

Fig. 9.3 Co-morbid conditions in our cohort of patients with DTD. *RAD* reactive airway disease, *LTB* laryngotracheobronchomalacia, *SDB* sleep disordered breathing

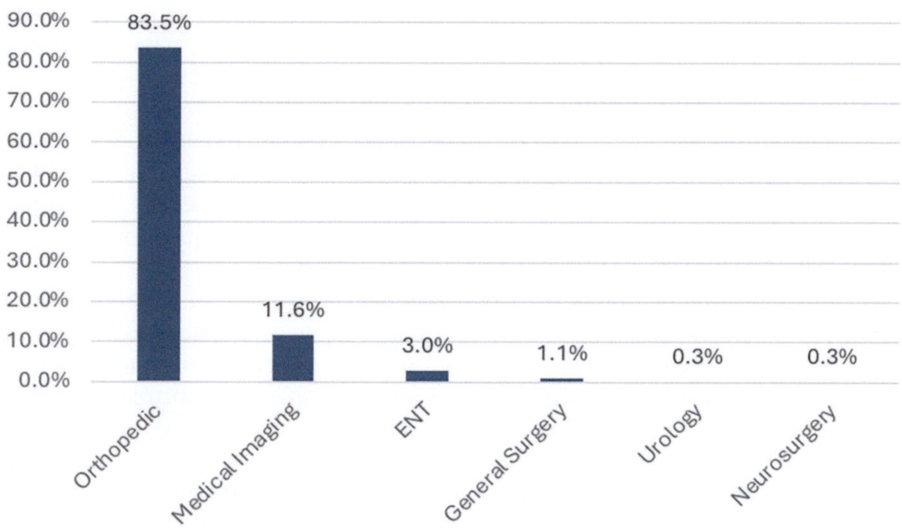

Fig. 9.4 Types of procedures the 44 patients underwent. ENT, otolaryngology

9.7 Anesthesia Care

There is no published information available to describe anesthetic care and related complications. Therefore, the information provided here is based on our experience at Nemours Children's Hospital, Delaware Valley. There were 174/267 surgeries managed with an endotracheal tube (ETT) for airway with no patient failing intubation. One patient had difficulty with bag mask ventilation and required an oral airway with two practitioners managing the airway. The intubation technique most used was direct laryngoscopy (56.9%). The next most common was video laryngoscopy (40.8%), for which GlideScope (Verathon, Bothell, WA) was used for 37.4% and C-MAC (Storz, El Segunado, CA) was used for 3.4%. Four patients required a flexible bronchoscopy aided intubation (FB); one of the patients who needed an FB intubation was managed by an otolaryngologist. The remaining procedures used either a laryngeal mask airway (17.6%) or a natural airway (4.1%) for airway management. Five patients had anesthetics managed with a tracheostomy; their ages ranged from 4 months to 12 years. Four of the 5 had tracheostomies at time of presentation to our institution. One patient had a tracheostomy placed at our institution due to spinal cord injury resulting in respiratory insufficiency/failure. There were 19 (7.1%) cases managed with a tracheostomy. Of note, while not all the same patients, 4 of the 5 patients had RAD and cleft palate. Sixteen cases did not have an airway documented (6.0%).

Three patients experienced stridor postoperatively: two in the recovery room and one at home a few hours after discharge. This last patient was seen in the emergency room. All were treated successfully with standard therapies. Difficulty with passing an ETT was encountered in three patients, requiring the ETT to be downsized.

Careful history and physical exam are necessary to safely provide anesthesia for patients with DTD. Review of systems should assess for laryngo-tracheo-broncho-malacia and SDB. In addition, it is important to establish whether the patient has any neurological symptoms related to possible spinal cord compression. Physical exam with close attention to airway assessment is necessary.

During induction of anesthesia and airway management, placing a rolled towel under the shoulders allows for neutral positing of the head and neck to avoid cord compression and ischemia. An inhalational or intravenous induction may be performed, keeping in mind that these patients are prone to airway obstruction due to micrognathia and laryngo-tracheo-broncho-malacia. In addition, the patients have restrictive lung disease due to a small rib cage and thoracic scoliosis, all of which may contribute to a smaller respiratory reserve and quicker oxygen desaturation if airway obstruction occurs. Conventional laryngoscopy may be adequate to intubate the trachea, but, as observed in our patients, difficult airway tools (namely video-laryngoscope or FB) may be needed for tracheal intubation.

In rhizomelic DTD, difficulty in placing the cuff of a sphygmomanometer may occur. Cuff width should be two-thirds of upper arm length with an inflatable bladder surrounding 40–80% of the arm circumference. With careful adjustment, it is possible to adequately measure blood pressure in most circumstances without resorting to invasive measures such as arterial pressure monitoring. Both venous and arterial access may be challenging as well, and use of ultrasound to obtain vascular access is highly recommended. Difficulty with positioning, especially for spine procedures, is common due to the shortened limbs with associated contractures and decreased range of motion. Particular attention should be paid to avoid pressure-related injuries, especially during long procedures such as spine fusion and or decompression.

Abnormalities of the vertebral structures compromising the spinal cord are commonly present, so DTD patients should be considered to have a "spine-at-risk" [10] unless cleared by their orthopedic surgeon or neurosurgeon. For any procedures under general anesthesia longer than 1 h, neuromonitoring should be considered [11]. Neuraxial anesthetics (spinal or epidural) may not only be difficult to perform, but they also may be contraindicated due to spinal cord compromise. If considering a peripheral nerve block, either as a single shot or placement of a catheter, the local anesthesia may need to be administered at the completion of surgery if neuromonitoring is being performed. This may be done successfully in collaboration with the surgeon's input.

9.8 Case Illustration

The patient initially presented to our institution at age 7 months for evaluation of cervical kyphosis with associated instability requiring a cervical collar to limit neck flexion. He was diagnosed with DTD in utero which was confirmed postnatally by genetic testing. Surgical history included a mandibular distraction during infancy for nasopharyngeal obstruction and two procedures for tympanostomy tubes. Initial MRI on presentation revealed a 93° cervical kyphosis with stenosis but no spinal cord changes. The degree of curvature was concerning, but because cervical kyphosis can improve over time with DTD, the decision was made to monitor the patient while continuing to use a cervical brace. The family was instructed to follow-up in 6 months.

At age 21 months, the patient returned for a follow-up visit, which was 8 months overdue. At this time, his mother reported that he was able to roll on the floor. He was not able to pull to sitting or standing, but he was able to sit if assisted. In addition, his mother stated that he was rolling less frequently than before. In the interim, he had cleft palate surgery at another pediatric institution in his home state. That surgery was associated with a tracheal intubation that took more than an hour and PICU admission for postoperative stridor.

An MRI with flexion and extension was ordered and performed at our institution to reassess the patient's cervical kyphosis. An inhalational induction with sevoflurane was used to induce general anesthesia. There was no difficulty with bag mask ventilation. Video laryngoscopy with the GlideScope Cobalt size 2 (Verathon) displayed a full view of the glottis, and a 3.0-cuffed ETT was placed easily. Anesthesia was maintained with sevoflurane in a mixture of air and oxygen without incident. At the conclusion of the MRI scan, the patient's trachea was extubated without incident. The patient was discharged the same day and

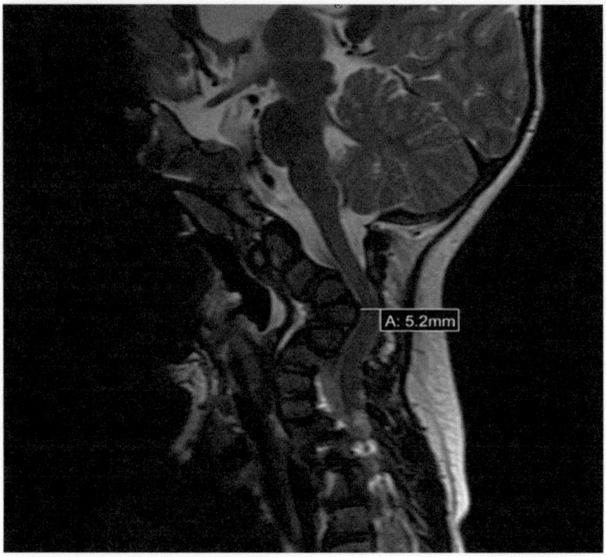

Fig. 9.5 Cervical kyphosis measuring at 97° at the level of C4 with complete effacement of the CSF

followed up in clinic the following day to review the results of the scan. MRI showed severe progressive cervical kyphosis measuring 97° with the apex of the curve at the level of the fourth cervical vertebrae (C4), complete effacement of the cerebrospinal fluid (CSF), surrounding the cord and minor cord compression at the level of C4 vertebral body (Fig. 9.5).

Due to progressive cervical kyphosis with spinal canal narrowing along with delays in motor skills, the patient underwent an anterior C4 vertebrectomy, C3–C5 instrumented fusion with cage placement, and a posterior C2–C7 fusion with iliac crest bone graft. After the vertebrectomy and anterior/posterior fusion, he was placed in a halo vest in slight extension to allow for healing.

For the cervical fusion, induction and intubation were performed in a manner similar to the technique used for the MRI. Peripheral IV access, an arterial line, and central venous access were obtained prior to incision. Due to the need for neuromonitoring, a total intravenous anesthetic (TIVA) using a combination of propofol and sufentanil was administered. The entire anesthetic and operative time were just shy of 11 h. The patient's trachea was extubated at the end of the case, and he was transferred to critical care for ongoing management.

The halo vest held the unstable spine in position and allowed time for the cervical fusion to heal and the bone to gain stability. This was removed after 4 months, and the patient was placed in a soft c-collar for 8 months after halo vest removal. Approximately 1 year after the initial surgery, all cervical spine precautions were cleared.

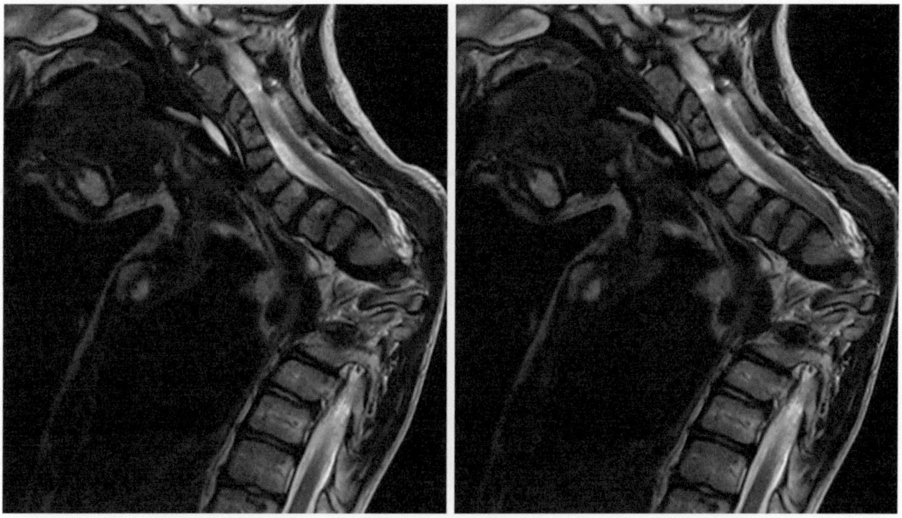

Fig. 9.6 MRI showing 85° thoracic kyphosis (both images) and post-surgical changes from C4 vertebrectomy, C3–C5 instrumented fusion with cage placement, and a posterior C2–C7 fusion

Over the course of the next several years, this patient did very well and was seen routinely for yearly checkups and health maintenance visits. From ages 6–9 years old, his mother noted progressive worsening of his back "hump" (upper thoracic kyphosis). Radiographic studies showed progressive upper thoracic deformity over those 3 years, measuring approximately 85° with apex at T5 (Fig. 9.6). Recommendations for a staged surgery with anterior disc release and halo gravity traction followed by placement of growing rods were made.

Initially, anterior thoracoscopic releases were done from T3–T8, which was followed by 4 weeks of halo gravity traction before implantation of growing rods. The anterior thoracoscopic release was done to straighten the thoracic kyphosis prior to the placement of growing rods. Figure 9.7 shows a photo of our patient in halo gravity traction. Surgically, the anterior thoracoscopic release involves division of the anterior longitudinal ligament and continued release until the annulus of the intervertebral disc is reached. This then allows the halo gravity traction to slowly assist in straightening the kyphosis.

An intravenous induction with propofol, remifentanil, and ketamine was used to induce general anesthesia. There was no difficulty with bag mask ventilation. Video laryngoscopy with a GlideScope size 2.5 blade (Verathon) was used for intubation of the trachea with a 4.0-cuffed ETT. After adequate IV access and an arterial line were secured, a TIVA anesthetic was used with a combination of propofol, remifentanil, and ketamine due to the need for neuromonitoring. The patient was positioned laterally for the surgical portion of the case. From start to finish, the case was about 5 h in length and uneventful. The patient's trachea was extubated, and he was taken to PACU without issues.

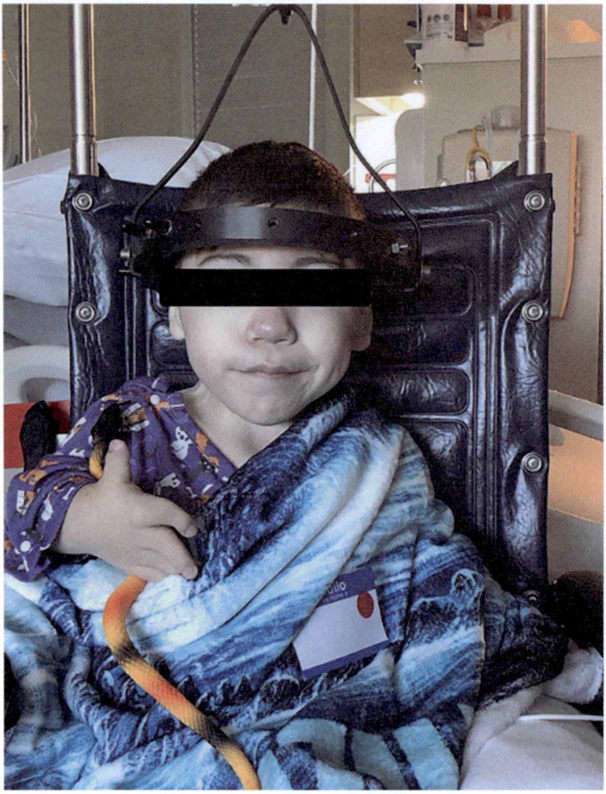

Fig. 9.7 Halo gravity traction after anterior thoracoscopic release and prior to placement of growing rods

After 4 weeks of halo gravity traction, our patient came back for placement of spinal growing rods. This type of rod needs to be adjusted in the office every 3–4 months. The adjustment is external and does not require anesthesia. The case was complex due to poor bone quality leading to difficulty with placement of spinal screws. The patient's dysplasia and non-ambulatory status could explain the overall poor bone quality. The case lasted for about 10 h, and a similar anesthetic approach was used as described for the anterior thoracoscopic release. Neuromonitoring remained stable throughout the case, the patient's trachea was extubated uneventfully, and he was taken to PACU. Figure 9.8 shows both AP and lateral views of our patient status post growing-rod placement. Our patient will continue to be followed and likely undergo his final spinal fusion in the upcoming years.

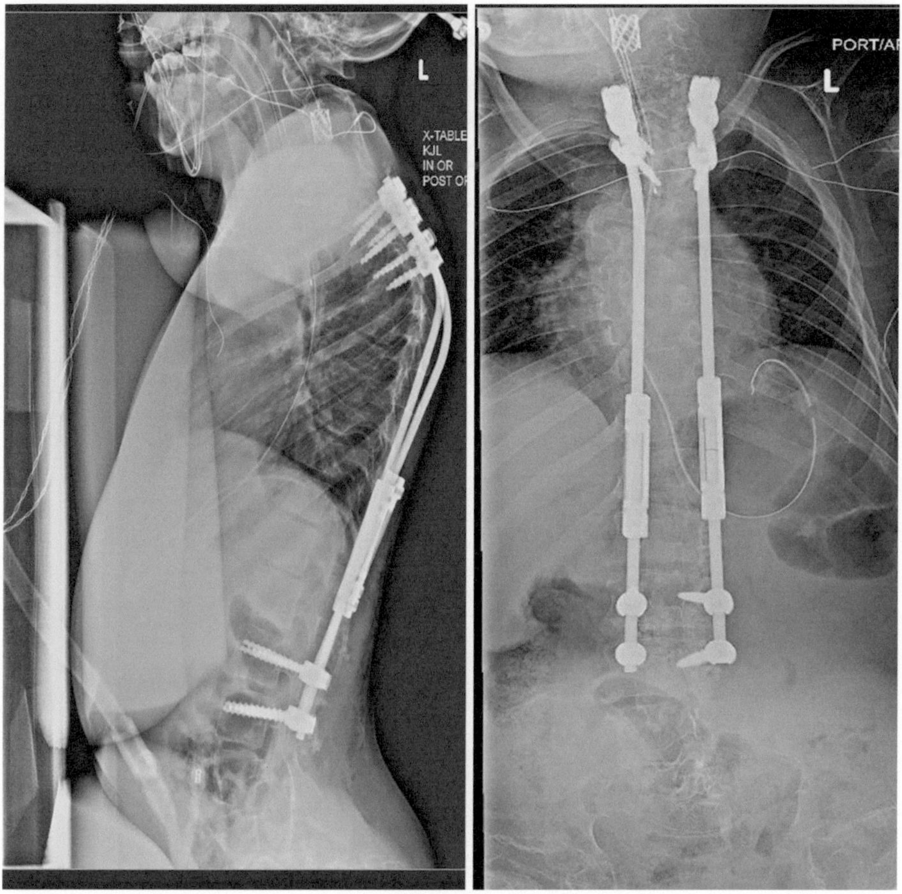

Fig. 9.8 Lateral (left) and AP (right) view of growing-rod hardware placed for thoracic kyphosis 4 weeks after halo gravity traction. Note the metal cage in the cervical region from the previous C4 vertebrectomy and cervical spinal fusion at age 21 months

References

1. Rossi A, Superti-Furga A. Mutations in the diastrophic dysplasia sulfate transporter (DTDST) gene (SLC26A2): 22 novel mutations, mutation review, associated skeletal phenotypes, and diagnostic relevance. Hum Mutat. 2001;17:159–71.
2. Parikh S, Batra P, Do TT. Diastrophic dysplasia. Medscape. 2021. https://emedicine.medscape.com/article/1257787-overview.
3. Harkonen H, Loid P, Makitie O. SLC26A2-associated diastrophic dysplasia and rMED-clinical features in affected Finnish children and review of the literature. Genes (Basel). 2021;12(5):714.
4. Honório JC, Bruns RF, Gründtner LF, Raskin S, Ferrari LP, Araujo Júnior E, et al. Diastrophic dysplasia: prenatal diagnosis and review of the literature. Sao Paulo Med J. 2013;131:127–32.

5. Unger S, Superti-Furga A. Diastrophic dysplasia. In: Adam MP, Everman DB, Mirzaa GM, Pagon RA, Wallace SE, Bean LJH, et al., editors. GeneReviews®. Seattle: University of Washington, Seattle; 1993.
6. Remes V, Marttinen E, Poussa M, Kaitila I, Peltonen J. Cervical kyphosis in diastrophic dysplasia. Spine (Phila Pa 1976). 1999;24:1990–5.
7. White KK, Sucato DJ. Spinal deformity in the skeletal dysplasias. Curr Opin Orthop. 2006;17:499–510.
8. Remes VM, Marttinen EJ, Poussa MS, Helenius IJ, Peltonen JI. Cervical spine in patients with diastrophic dysplasia--radiographic findings in 122 patients. Pediatr Radiol. 2002;32:621–8.
9. Remes V, Tervahartiala P, Poussa M, Peltonen J. Thoracic and lumbar spine in diastrophic dysplasia: a clinical and magnetic resonance imaging analysis. Spine (Phila Pa 1976). 2001;26:187–95.
10. White KK, Bompadre V, Goldberg MJ, Bober MB, Cho TJ, Hoover-Fong JE, et al. Best practices in peri-operative management of patients with skeletal dysplasias. Am J Med Genet A. 2017;173:2584–95.
11. White KK, Bober MB, Cho TJ, Goldberg MJ, Hoover-Fong J, Irving M, et al. Best practice guidelines for management of spinal disorders in skeletal dysplasia. Orphanet J Rare Dis. 2020;15:161.

Larsen Syndrome

10

Connie J. Lin, Manish Purohit, and Lauren W. Averill

10.1 Introduction

Larsen syndrome is a rare form of skeletal dysplasia occurring in approximately 1 in 100,000 births [1, 2]. Initially described by Dr. Loren Larsen in 1950, it is characterized by flat facies and multiple major joint dislocations with variable severity and presentation [3]. In addition to the skeletal manifestations, anomalies of the airway, heart, and neuromuscular system may also be present.

The syndrome is caused by mutations in the filamin B (*FLNB*) gene, resulting in an abnormal filamin B protein. The resulting altered filamin B protein is unable to link actin filaments and communicate with cellular membranes, preventing the normal protein complex required for creating and maintaining cellular structure [4–6]. With this loss of integrity, chondrocytes, osteoblasts, and fibroblasts are unable to orient appropriately, resulting in the skeletal abnormalities associated with Larsen syndrome [4].

Patients with Larsen syndrome require multiple surgical and nonsurgical procedures during their lifetime and present with evolving anesthetic and perioperative challenges secondary to co-morbidities and characteristics associated with this rare dysplasia [7, 8].

C. J. Lin (✉) · M. Purohit
Anesthesiology and Pediatrics, Sidney Kimmel Medical College,
Thomas Jefferson University, Philadelphia, Pennsylvania, USA

Department of Anesthesiology and Perioperative Medicine, Nemours Children's Hospital,
Wilmington, Delaware, USA
e-mail: Connie.Lin@nemours.org; Manish.purohit@nemours.org

L. W. Averill
Radiology and Pediatrics, Sidney Kimmel Medical College, Thomas Jefferson University,
Philadelphia, Pennsylvania, USA

Department of Radiology, Nemours Children's Hospital, Wilmington, Delaware, USA
e-mail: Lauren.averill@nemours.org

M. C. Theroux (ed.), *Anesthesia Care of Patients with Skeletal Dysplasia*,
https://doi.org/10.1007/978-3-031-92501-6_10

10.2 Genetics

Historically, Larsen syndrome is considered to be an autosomal dominant syndrome, with mutations in the *FLNB* gene on chromosome 3p14.3 occurring in approximately 1 in 100,000 people [9–12]. The possibility of an autosomal recessive form of Larsen syndrome has come under some debate as "Larsen-like" disease processes with an autosomal recessive form of transmission have been labeled as Larsen syndrome. These autosomal recessive syndromes tend to be more severe in their presentation and involve multiple organ systems while still presenting with the basic Larsen phenotype. Despite these clinical presentations being phenotypically discrete from the autosomal dominant syndrome, it is accepted that an autosomal recessive form of Larsen syndrome caused by either a missense mutation or mosaicism affecting the *FLNB* gene cannot be ruled out at this time [13–16].

Of note, mutations in the *FLNB* gene may cause other syndromes distinct from and often more severe than Larsen syndrome, including atelosteogenesis types I and III as well as boomerang dysplasia [17, 18]. These syndromes, unlike Larsen, may cause perinatal lethality.

10.3 Phenotypes

Based on the original description by Larsen in 1950, patients present with varying symptoms secondary to chondrocytic and bone malformations throughout the entire skeleton (Table 10.1). Intellectual function is usually unaffected. Patients may present or be born with dislocations in the hips, knees, shoulders, or elbow joints as well as club feet. Extraneous small bones in their limbs, such as the wrists and ankles, may lead to functional issues. Small digits also may be affected; the tips of the

Table 10.1 Classic manifestations of Larsen syndrome [3]

Joint dislocations	Hip
	Knees
	Shoulder
	Elbow
Foot deformities	Clubfoot
	Supernumerary tarsal bones
Hand deformities	Spatula-shaped or cylindrical fingers
Spine abnormalities	Scoliosis
	Cervical kyphosis
Facial anomalies	Prominent forehead hypertelorism
	Depressed nasal bridge
	Malar flattening
	Cleft palate
	Bifid uvula
Deafness	
Tracheomalacia	

Table 10.2 Subsequent findings by organ system from a recent publication [20]

Joint dislocations	Elbows
	Wrists
	Hips
	Knees
Spine anomalies	Atlantoaxial instability
	Vertebral segmentation anomalies
	Hemivertebrae
	Scoliosis
	Cervical kyphosis
	Thoracolumbar kyphosis
Facial anomalies	Saddle nose
	Malar hypoplasia
	Frontal bossing
	Fusion of frontal and parietal bones
	Low-set ears
	Cleft palate
	Short, broad neck with redundant tissue
Cardiac anomalies	ASD/VSD
	Mitral/aortic/tricuspid valvulopathies
	Aortic root enlargement
	Subaortic stenosis
	Patent ductus arterioles
	Cardiomyopathy
Respiratory anomalies	Laryngomalacia
	Pulmonary hypoplasia
	Restrictive lung disease
	Subglottic stenosis
	Barrel chest
	Pectus excavatum
Deafness	

thumbs might look blunt and square-shaped or spatulate. Conditions that later develop include kyphosis or scoliosis with spinal segmentation and possible cervical myelopathy. Atlantoaxial instability and sub-axial cervical kyphosis may also develop, leading to spinal cord compression and impaired breathing [3].

Involvement of the skull and facial bones can present as frontal bossing, midface hypoplasia, ocular hypertelorism, and sometimes cleft palate, but not cleft lip. Variable hearing loss may also be present due to malformation of ossicles in the ear [3, 19].

As noted above, multiple syndromes that share similar physical characteristics with Larsen syndrome were classified as such but often were not caused by an abnormality in the specific exons of the filamin B gene on chromosome 3p14.3. This inconsistency has led to a wide variability in the phenotype of Larsen syndrome and the subsequent diagnosis of autosomal recessive forms. These autosomal recessive conditions, now also classified as Larsen syndrome, are generally considered more severe and present with additional organ system involvement as described in Table 10.2 [15, 21]. At this time, there is no clear consensus delineating these autosomal recessive forms from the autosomal dominant counterpart [22, 23].

10.4 Radiological Findings

Radiological findings of concern to the anesthesiologist are the cervical spine related abnormalities, such as kyphotic curve of or instability of the cervical spine at any level. Typical findings in a patient with Larsen syndrome are presented here (Fig. 10.1).

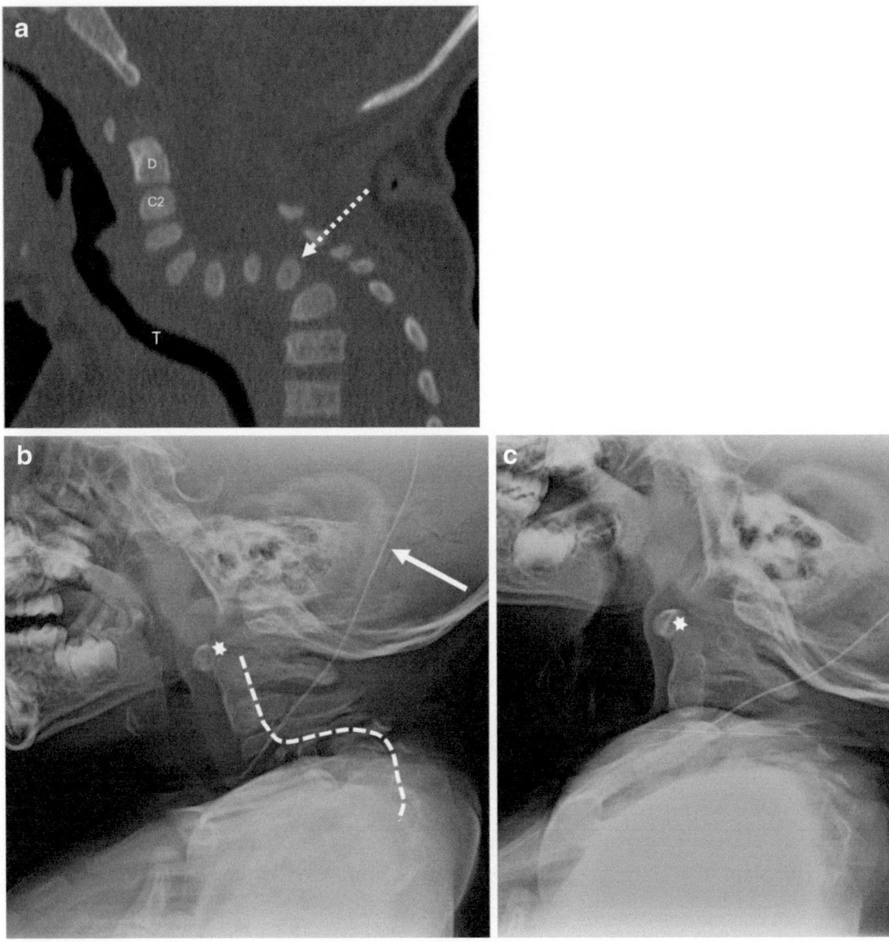

Fig. 10.1 Seven-year-old boy with Larsen syndrome and the typical appearance of the cervical spine. Sagittal reconstruction of a CT of the cervical spine (**a**) performed at age 6 months shows severe kyphosis centered at C7 (dashed arrow) and additional upper cervical lordosis (D = dens, C2 = body of C2). Trachea (T) is normal in caliber, draping over the cervical spine. Lateral cervical spine x-rays of the above patient in neutral (**b**) and extension (**c**) performed at age 3 years, 11 months show progressive cervical lordosis and cervicothoracic kyphosis (dashed line highlighting posterior margin of vertebral bodies). The atlanto-dens interval (star) is normal throughout the range of motion. Ventricular shunt tubing projects over the image (arrow). Sagittal reconstruction of a CT of the cervical spine of the same patient (**d**) performed at age 8 years shows appearance of the spine after cervicothoracic spinal fusion, with large bony fusion mass (straight arrows) and hardware (curved arrow). Trachea (T) remains normal in caliber, draping over the cervical spine

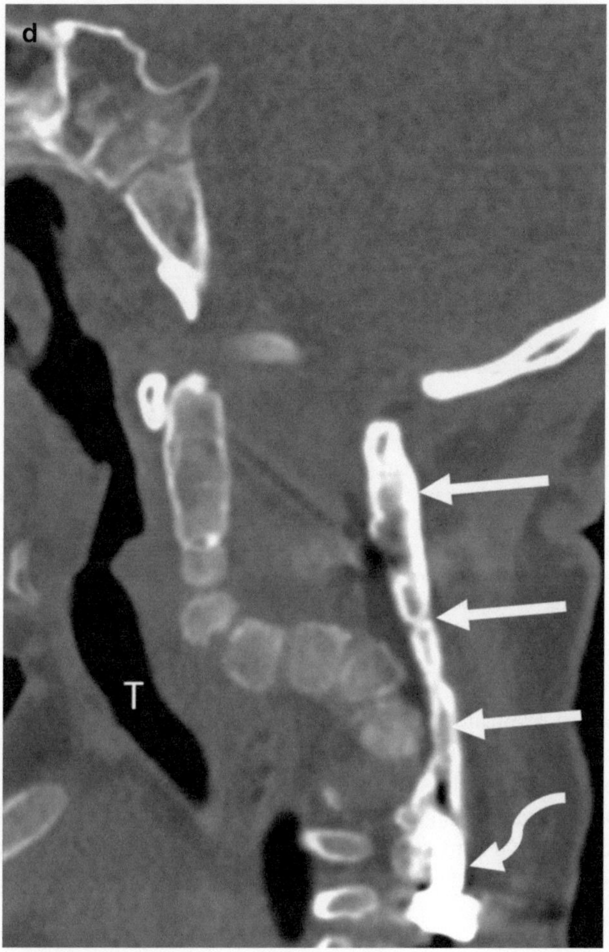

Fig. 10.1 (continued)

10.5 Co-morbid Conditions

Co-morbidities of Larsen syndrome vary greatly due to the heterogeneic inconsistency associated with the classification of the patients. The organ systems most commonly affecting anesthetic management are the airway, respiratory and cardiac ac systems as noted in Table 10.2.

Cervical spine instability is a hallmark of Larsen syndrome, with progressive kyphosis causing spinal cord compression at the cervical level, which, if untreated, may lead to cervical myelopathy and paralysis. As the cervical spine abnormalities are corrected surgically with decompression and fusion, airway management may become progressively more difficult due to increased lack of mobility and shortness of the head and neck (Fig. 10.2).

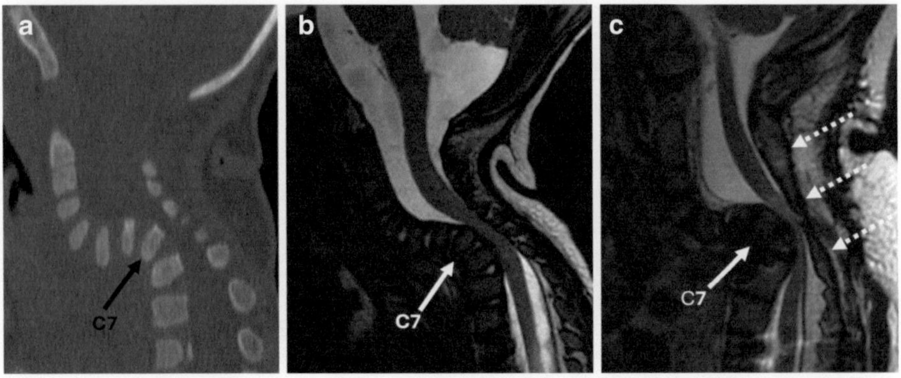

Fig. 10.2 Boy with Larsen syndrome with severe lower cervical kyphosis and myelopathy. At age 6 months, sagittal CT of the cervical spine (**a**) shows severe kyphosis centered at C7, with narrowing of the spinal canal. Spinal canal above and below the kyphosis is capacious. Sagittal T2-weighted MR image of the cervical spine (**b**) at age 8 months shows similar findings, with complete effacement of cerebrospinal fluid at the cervicothoracic junction. Sagittal T2-weighted MR image of the cervical spine (**c**) at age 9 years shows cervicothoracic spinal fusion with mature bony mass posteriorly (dashed arrows) and greater than 90° kyphosis at C7. Dark signal in the C6–T2 vertebrae is due to artifact from hardware. Note spinal cord thinning at the cervicothoracic junction

Pulmonary and respiratory function may be affected due to physical obstruction from midface hypoplasia causing sleep apnea, tracheobronchomalacia causing chronic stridor, or from scoliosis causing a restrictive lung disease (as is common in many forms of skeletal dysplasia). Tracheobronchomalacia and scoliosis can lead to a combined obstructive and restrictive lung disease further complicating patient well-being, especially when cardiac anomalies may be involved as well [7, 8].

10.6 Anesthetic Care

10.6.1 Preoperative Management

Larsen syndrome is a progressive condition with evolving comorbidities requiring multiple surgeries. Predominant surgical procedures include orthopedic procedures to correct the various musculoskeletal abnormalities. Preanesthetic evaluation should consist of a thorough history and physical with particular attention applied to prior anesthetic management and postoperative course.

Cervical spine abnormalities are often severe and frequently complicate airway management [8]. A review of prior anesthetics with special attention to the airway management is essential. It will be unusual to have a patient with Larsen syndrome present to the anesthesia provider without imaging studies to evaluate the cervical spine.

Flexion/extension studies of cervical spine are very informative and are taken as plain x-rays and as MRI [24]. These studies enable us to evaluate any instability present at any of the levels of cervical spine. Modalities such as this require close coordination with the radiologist to facilitate timely interpretation of images. Besides concerns regarding cervical spine abnormalities requiring special considerations for airway management and endotracheal intubation, other features contributing to a difficult airway are midface hypoplasia and maxilla-fascial abnormalities.

Respiratory system abnormalities are a hallmark of Larsen syndrome, ranging from subglottic stenosis to laryngo-tracheo-malacia. When these pulmonary anomalies are associated with severe scoliosis, patients may develop a thoracic insufficiency syndrome. Preoperative imaging of the thoracic and lumbar spine will allow for evaluation of the severity of the scoliosis and for appropriate planning for positioning and airway management. Otolaryngological evaluation may further demonstrate the caliber of the airway, the severity of subglottic stenosis, and the resulting airway compromise.

Cardiac defects are varied in Larsen syndrome and ultimately could contribute to thoracic insufficiency syndrome, compounding an already compromised respiratory system [25]. A cardiology consult is recommended, and electrocardiogram and echocardiogram at the discretion of a cardiologist will reveal current cardiac status. In some cases, consultation with pediatric cardiac anesthesiology may be warranted.

10.6.2 Intraoperative Management

Induction of patients with Larsen syndrome may be performed via an intravenous or inhalation technique. While an intravenous induction may be associated with fewer potential complications and allow rescue medications to be administered more easily, should the need arise, an inhalation technique is also possible. However, it must be undertaken slowly and deliberately, assessing the airway as the patient is induced. Medications commonly used for induction include propofol, ketamine, fentanyl, and remifentanil (often used to facilitate intubation in lieu of muscle relaxant[1]) depending upon the anticipated degree of difficulty with the airway management. Maintenance of anesthesia is accomplished using propofol and remifentanil (current mainstay of the infusions). Long-acting narcotic infusions are avoided to allow for more predictable titration and to facilitate an awake extubation at end of the surgical procedure.

During induction and airway manipulation, patients should be maintained in a neutral to slightly hyperextended position such that the tragus of the ear is at the level of the clavicle. A small shoulder roll is placed to achieve this position, which not only optimizes the alignment of the cervical spine but also facilitates airway management.[2] We have experienced change in evoked potentials following a direct laryngoscopy with Mallampati class 1 airway [8]. Our recommendation is that the patient should be positioned with head and neck in a neutral position as previously

[1] Please refer to Chap. 16: Anesthetic Management for Spine Fusion

[2] Please refer to Chap. 16: Anesthetic Management of Spine Fusion

described. Tracheal intubation should be accomplished using video laryngoscopy or a flexible bronchoscope.

As is common with many forms of skeletal dysplasia, it is necessary to downsize the endotracheal tube from what may be considered appropriate for the patient's age [8]. The caliber of the tracheal lumen often will not be sufficient to allow the tube size deemed appropriate for age. As these patients' body mass is considerably less than that of similarly aged patients of average stature, the smaller endotracheal tubes are often quite sufficient to meet the ventilatory demand. Further downsizing of the endotracheal tube may be required in the presence of previously undiagnosed subglottic stenosis.

For extensive surgical procedures such as spine fusion, fluid management should take into consideration the necessity of extubating the trachea without incurring undue edema of the airway structures. It may be fair to give priority to the concerns regarding edema to the airway when correcting blood loss such that there is minimal airway swelling, thus enabling the trachea to be extubated at the end of the procedure. Venous and arterial access placement may be at the discretion of the anesthesiologist with a goal of maintaining mean arterial pressures of 60–70 mmHg or higher to optimize perfusion to the spinal cord. To achieve this, it may be necessary to use vasopressors.

Neuromonitoring should be an integral part of the overall anesthetic plan for patients with Larsen syndrome. They should be considered as having a 'spine at risk'; therefore, any anesthetic procedure that requires longer than 60 min should have neuromonitoring in place. When no neuromonitoring is possible (as in an MRI suite), it is important to have a discussion with the orthopedic team, radiologist, and radiology technician to determine the safest position for the patient (lateral decubitus is usually preferred). Depending on the circumstances, it may be necessary to limit the amount of time that patient is under deep sedation or general anesthesia. Patients with Larsen syndrome may present with kyphosis and other bony abnormalities that require careful positioning and padding to prevent pressure-related injury. We recommend having a discussion with the patient or parents to determine which position the patient sleeps in most comfortably. This often gives the anesthesia providers valuable information about how the patient keeps his/her head and neck and which position compromises their spinal cord the least. A discussion with neuromonitoring personnel is invaluable in optimizing the intraoperative positioning as well.

10.7 Postoperative Care

Following extubation, increased respiratory effort from hypotonia and tracheo-broncho-malacia may arise, requiring ventilatory or oxygen support. Stridor, regardless of the presence of subglottic stenosis, can be treated with racemic epinephrine and intravenous steroids. Inhaled bronchodilators such as albuterol may reverse any bronchospasm that is present sub-clinically (may not be audible when auscultating). Frequently, we employ noninvasive techniques such as humidified high-flow nasal cannula (Vapotherm, Exter, New Hampshire) and CPAP/BiPAP (especially when patients have been using it prior to surgery) to stabilize the patient in the immediate postoperative period. This noninvasive respiratory support further allows an increased margin of safety during the time the anesthesia provider is treating the patient for pain and or spasm with medications that cause respiratory depression. Humidified high-flow nasal cannula is particularly effective in decreasing the work of breathing and adds a layer of safety to the immediate postoperative period for these patients, most of whom suffer from a restrictive airway disease.

10.7.1 Pain Management

Pain management starts in the operating room before the patient emerges from anesthesia. In patients with Larsen syndrome, as in many other patients with skeletal dysplasia who have a history of a difficult airway, long-acting narcotics should be minimized to ensure an awake and safe extubation in the operating room. Postoperatively, the commonly used medications for pain management are scheduled nonsteroidal anti-inflammatory drugs and acetaminophen. First doses of these medications are typically given prior to emergence. Judicious use of benzodiazepines for muscle spasm may be prescribed but with caution. Agonist/antagonist analgesic medications such as nalbuphine are an excellent choice for further pain control with minimal effect on respiratory drive in these patients. Whenever possible, regional anesthesia techniques utilizing peripheral nerve blocks should be employed while avoiding the neuraxial technique due to the prevalence of spinal abnormalities. Because of the need for neuromonitoring, regional anesthesia should only be employed at the completion of the surgical procedure and the neuromonitoring. If necessary, a physical examination may be performed to assess the integrity of the spinal cord prior to initiating regional blocks using local anesthetics. Additives to the local anesthetic solution are generally not used for such blocks due to instances of prolonged block.

10.8 Case Illustration

A 15-year-old girl (weight, 51.4 kg; height, 132.8 cm) with Larsen syndrome and severe kyphoscoliosis presented for posterior spinal fusion. She was the product of a normal spontaneous vaginal delivery and spent 5 days in the neonatal intensive care unit for stridor due to laryngomalacia. She had a history of bilateral dislocated knees and hips as well as clubfeet, diagnosed at birth, and had been treated with casts until age 6 months. At age 12 months, her neck was assessed at an outside facility by a physician who did not find any cervical spine instability. She also had a history of auditory deficits in the right ear and used binaural hearing aids.

At age 28 months, she underwent posterior arthrodesis of the upper thoracic spine and lumbar spine for severe kyphoscoliosis at an outside pediatric tertiary care center. Growing rods were placed during this surgery, and she returned every 6 months for rod expansion and lengthening.

At age 5 years, she underwent bilateral posterior release of clubfeet and bilateral hemiepiphysiodesis with 8-plates at her home institution. At age 6 years, she underwent bilateral anterior distal femoral hemiepiphysiodesis and bilateral Achilles tendon lengthening. However, she continued to have left-knee flexion contracture. As a result, she came to our institution at age 8 years for removal of 8-plates, left-knee capsular release, and left distal femoral extension osteotomy under general anesthesia with an endotracheal tube. For this procedure, the patient underwent an uneventful inhalation induction and intravenous line placement. Intubation was easily attempted with direct laryngoscopy and a 5.5 cuffed endotracheal tube but did not have a leak. As a result, the endotracheal tube was downsized to 4.5 tube with an appropriate leak at less than 25 mmHg. The procedure was completed uneventfully, and the patient was extubated awake and taken to the post-anesthesia care unit for recovery.

At age 11, the parents transferred the care of the patient to our institution. She subsequently underwent rod removal and hardware replacement of her spine for a wound infection. For this procedure, she again had an inhalational induction followed by intravenous line placement. Her airway was secured with direct laryngoscopy with a Miller #2 blade and a 5.5 cuffed endotracheal tube. Total intravenous anesthesia was administered and maintained with propofol and fentanyl infusions for neuromonitoring. The anesthesia and surgery proceeded uneventfully, and the patient was subsequently extubated awake and transferred to the post-anesthesia care unit.

She continued to have spine lengthening every 6 months until age 15 years, at which time she was scheduled for a definitive spinal fusion. At this time, she was also noted to have significant cervical kyphosis. Imaging studies (MRI and CT) of her cervical, thoracic, and lumbar spine showed severe scoliosis with focal kyphosis at C1–C2 vertebrate levels with focal narrowing of the spinal canal (Fig. 10.3c, d for the CT images, MRI images were not available to us). The MRI also showed kyphosis in the upper thoracic spine and focal narrowing at the cervicothoracic junction. Partial effacement of CSF and mild flattening of the spinal cord were also noted at the cervicothoracic junction.

The patient presented for removal of instrumentation and halo placement at age 15 years. Anesthesia was induced with nitrous oxide and a small dose of sevoflurane to facilitate placement of an intravenous access and to proceed with full induction of anesthesia. Her neck was maintained in a neutral position. Anesthesia was deepened with a bolus of propofol 100 mg and remifentanil 40 mcg. The trachea was intubated using a video-laryngoscope and a 5.5 cuffed endotracheal tube. An arterial line was placed for hemodynamic monitoring and drawing necessary blood samples for lab work. Total intravenous anesthesia was maintained with a propofol and remifentanil infusion to facilitate neuromonitoring of evoked potentials. The patient was placed in the prone position for removal of instrumentation and then turned supine for the placement of halo traction. The surgery and anesthetic proceeded uneventfully and with stable evoked potentials. The patient was awakened

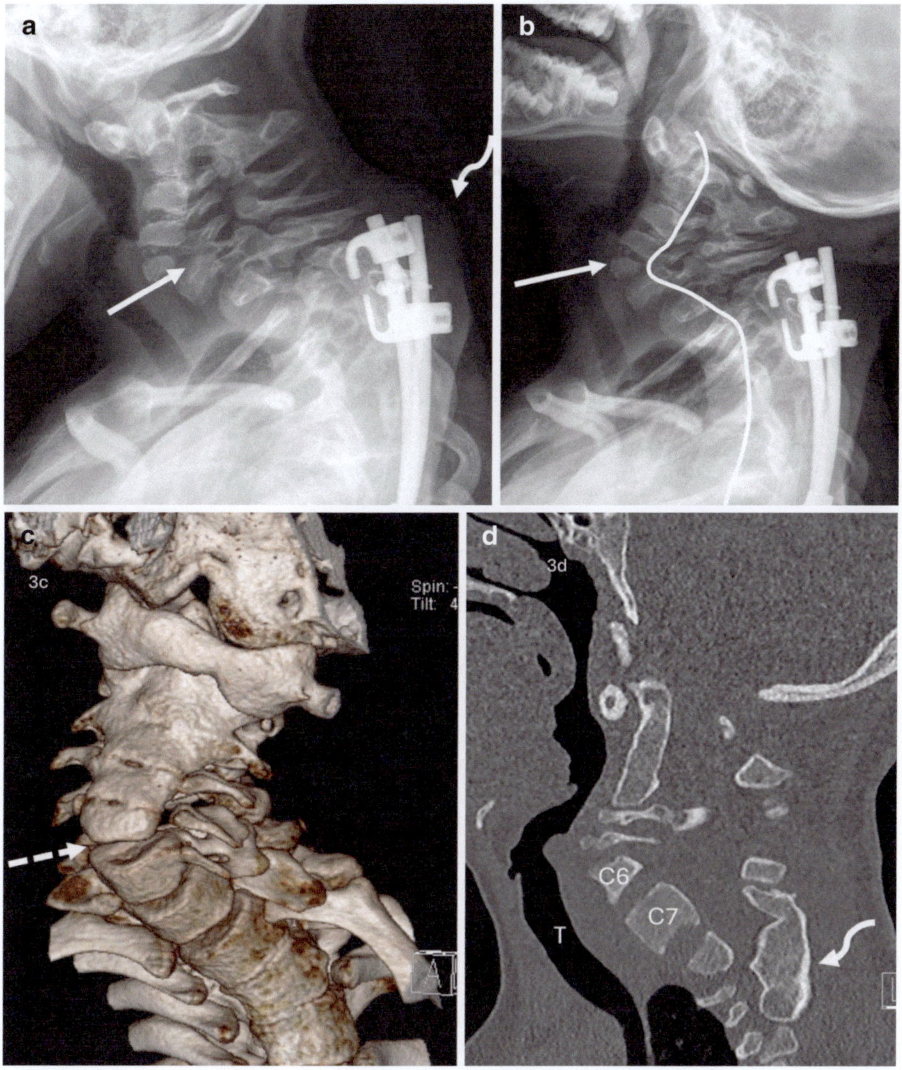

Fig. 10.3 Girl with Larsen syndrome and severe lordosis of the lower cervical spine. (**a**) Flexion and (**b**) extension lateral x-rays of the cervical spine show severe lordosis of the lower cervical spine centered at the hypoplastic C5 vertebra (arrow). Dashed line marks the posterior margin of the vertebral bodies, showing greater than 90° lordosis in extension. She previously underwent thoracolumbar spinal fusion (curved arrow). (**c**) 3D surface-rendered image from CT of the cervical spine performed at age 15 years shows bony fusion of the upper cervical spine and severe lordosis at C5–6 (dashed arrow). (**d**) Sagittal reconstruction from the same CT scan shows normal caliber of the trachea (T) as it drapes over the cervical spine. Bony anatomy is not displayed fully because the spine is not in the same sagittal plane as the trachea. C6 = C6 vertebra, C7 = C7 vertebra, curved arrow = lower cervical/upper thoracic bony fusion. (**e**) Lateral flexion and (**f**) extension x-rays of the cervical spine at age 20 years show anterolisthesis of C5 (arrow) and lordosis at C5–C6. Spinal fusion hardware has been extended superiorly to the C6 level (curved arrow). There is very little motion between the fused upper cervical spine and fused lower cervical/thoracic spine segments

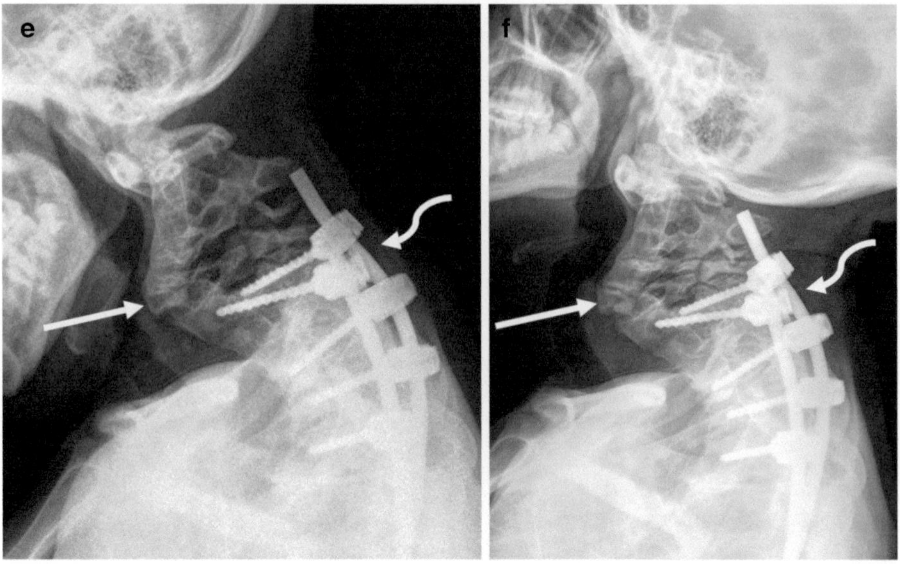

Fig. 10.3 (continued)

and extubated at the end of the procedure without difficulty and transferred to the recovery room.

Subsequently, she remained in the hospital and was followed by the pain service. Pain control was achieved with a combination of acetaminophen, nonsteroidal anti-inflammatory agents, diazepam, oxycodone, and morphine only for break-through pain. Repeat imaging of her spine after halo traction showed severe scoliosis with focal kyphosis at C1–2, focal lordosis in the mid-to-lower cervical spine, and additional kyphosis in the upper thoracic spine. Focal narrowing of the spinal canal at C1–2 and again at the cervicothoracic junction also were noted. There was partial effacement of cerebrospinal fluid and mild flattening of the spinal cord at the cervicothoracic junction.

Four weeks after placement of halo traction, the patient presented for posterior spinal fusion. Inhalation induction was performed uneventfully with a combination of oxygen, nitrous oxide and sevoflurane. The halo and traction remained in place for inhalation induction, and the head and neck were maintained in a neutral position. After IV line placement, the traction was removed with halo in situ for intubation with the neck remaining in neutral position. The airway again was secured with a video-laryngoscope and 5.5 cuffed ETT. A second IV and an arterial line were placed, similar to the prior surgery for hemodynamic monitoring and intraoperative lab draws. Baseline evoked potentials were obtained, and the patient then was placed in the prone position without difficulties. Anesthesia was maintained with a propofol and remifentanil infusion, and the depth of anesthesia was confirmed periodically with neuromonitoring. Tranexamic acid bolus and infusion were administered to reduce risk of bleeding.

The patient remained hemodynamically stable throughout the intraoperative course without any changes to baseline evoked potentials and did not require vasopressors. She required transfusion of 1 unit of packed red blood cells, 1 unit for fresh frozen plasma, and 500 ml of salvaged red blood cells during the procedure for an intraoperative blood loss of 1800 ml. The patient was extubated awake at the end of the procedure and was taken to the recovery room. She was admitted to the orthopedic service in the inpatient floor and subsequently discharged home on postoperative day 6.

Six months after surgery, the patient was doing well and walking unassisted with a walker as well as participating in girl scouts, band, and swimming. At 2-year follow-up, the patient continues to do well and is attending college.

This case highlights the classic phenotypical presentations associated with Larsen syndrome and the ongoing care provided from an early age. Our patient was born with multiple manifestations of Larsen syndrome, including abnormalities of the skeletal system, hearing loss, and laryngomalacia. Fortunately, her laryngomalacia resolved on its own without any interventions, and her hearing loss was treated with biaural hearing aids. Her initial skeletal dysplasia included hip and knee dislocations and club feet, which required casts and molded ankle-foot orthoses and then subsequent surgeries. Although her cervical spine was initially stable, she developed cervical instability and significant kyphosis of the cervicothoracic spine subsequently, requiring correction. She also had growing rods, which, at age 15 years, needed to be removed for her final posterior spine fusion. However, to facilitate correction of her kyphoscoliosis, she was placed in halo traction for 4 weeks prior to the posterior spinal fusion. She tolerated the various surgical procedures successfully and emerged as a fully functioning individual.

This case illustrates the long path of surgical procedures a typical patient with Larsen syndrome travels and the importance of multidisciplinary care needed to help them through it all successfully.

References

1. Kodra N, Diamonstein C, Hauser NS. A case study of atypical Larsen syndrome with absent hallmark joint dislocations. Mol Genet Genomic Med. 2019;7:e648.
2. Zhang D, Herring JA, Swaney SS, McClendon TB, Gao X, Browne RH, et al. Mutations responsible for Larsen syndrome cluster in the FLNB protein. J Med Genet. 2006;43:e24.
3. Larsen LJ, Schottstaedt ER, Bost FC. Multiple congenital dislocations associated with characteristic facial abnormality. J Pediatr. 1950;37:574–81.
4. Xu Q, Wu N, Cui L, Qiu G. Filamin B: the next hotspot in skeletal research? J Genet Genomics. 2017;44(7):335–42.
5. Robertson S. FLNB disorders. In: Adam MP, Mirzaa GM, Pagon RA, Wallace SE, Bean LJH, Gripp KW, et al., editors. GeneReviews (Internet). Seattle: University of Washington, Seattle; 2020.
6. McKay SD, Al-Omari A, Tomlinson LA, Dormans JP. Review of cervical spine anomalies in genetic syndromes. Spine (Phil PA 1976). 2012;37:E269–77.
7. Tobias JD. Anesthetic implications of Larsen syndrome. J Clin Anesth. 1996;8:255–7.

8. Malik P, Choudhry DK. Larsen syndrome and its anaesthetic considerations. Paediatr Anaesth. 2002;12(7):632–6.
9. Latta RJ, Graham CB, Aase J, Scham SM, Smith DW. Larsen's syndrome: a skeletal dysplasia with multiple joint dislocations and unusual facies. J Pediatr. 1971;78:291–8.
10. Alembik Y, Stoll C, Messer J. On the phenotypic overlap between "severe" oto-palato digital type II syndrome and Larsen syndrome. Variable manifestation of a single autosomal dominant gene. Genet Couns. 1997;8:133–7.
11. Stanley D, Seymour N. The Larsen syndrome occurring in four generations of one family. Int Orthop. 1985;8:267–72.
12. Vujic M, Hallstensson K, Wahlstrom J, Lundberg A, Langmaack C, Martinson T. Localization of a gene for autosomal dominant Larsen syndrome to chromosome region 3p21.1-14.1 in the proximity of, but distinct from, the COL7A1 locus. Am J Hum Genet. 1995;57:1104–13.
13. Oki T, Terashima Y, Murachi S, Nogami H. Clinical features and treatment of joint dislocations in Larsen's syndrome. Report of three cases in one family. Clin Orthop Relat Res. 1976;119:206–10.
14. Rochelson B, Petrikovsky B, Shmoys S. Prenatal diagnosis and obstetric management of Larsen syndrome. Obstet Gynecol. 1993;81:845–7.
15. Strisciuglio P, Sebastio G, Andria G, Maione S, Raia V. Severe cardiac anomalies in sibs with Larsen syndrome. J Med Genet. 1983;20:422–4.
16. Topley JM, Varady E, Lestringant GG. Larsen syndrome in siblings with consanguineous parents. Clin Dysmorphol. 1994;3:263–5.
17. Hunter AG, Carpenter BF. Atelosteogenesis I and boomerang dysplasia: a question of nosology. Clin Genet. 1991;39:471–80.
18. Bicknell LS, Morgan T, Bonafé L, Wessels MW, Bialer MG, Willems PJ, et al. Mutations in FLNB cause boomerang dysplasia. J Med Genet. 2005;42:e43.
19. Marques LHS, Martins DV, Juares GL, Lorenzetti FTM, Monsanto RDC. Otologic manifestations of Larsen syndrome. Int J Pediatr Otorhinolaryngol. 2017;101:223–9.
20. Siafaka A, Angelis S, Piagkou M, Apostopoulos A, Troupis T, Filippou D. Larsen syndrome and associated spinal disorders. Cureus. 2023;15:e41655.
21. Cetta G, Lenzi L, Ruggeri A, Tenni R, Boni M. Biochemical and structural abnormalities of the connective tissue in Larsen's syndrome. Int Orthop. 1979;3:47–53.
22. Steel HH, Kohl EJ. Multiple congenital dislocations associated with other skeletal anomalies (Larsen's syndrome) in three siblings. J Bone Joint Surg Am. 1972;54:75–82.
23. Mostello D, Hoechstetter L, Bendon RW, Dignan PS, Oestreich AE, Siddiqi TA. Prenatal diagnosis of recurrent Larsen syndrome: further definition of a lethal variant. Prenat Diagn. 1991;11:215–25.
24. Mackenzie WG, Dhawale AA, Demczko MM, Ditro C, Rogers KJ, Bober MB, et al. Flexion-extension cervical spine MRI in children with skeletal dysplasia: is it safe and effective? J Pediatr Orthop. 2013;33:91–8.
25. Campbell RM Jr. Spine deformities in rare congenital syndromes: clinical issues. Spine (Phila Pa 1976). 2009;34:1815–27.

Metatropic Dysplasia

11

Matthew McCoy, Melinda Brown, Lauren W. Averill, and Mary C. Theroux

11.1 Introduction

Metatropic dysplasia (MD), first described in 1966 [1], is an atypical form of skeletal dysplasia and 1 of more than 450 identified genetic skeletal disorders [2]. Metatropic dysplasia is caused by a mutation in the *TRPV4* gene and is characterized by short stature, short limbs, and severe kyphoscoliosis [3]. Although three distinct inheritance patterns have been described [4] (both lethal and non-lethal autosomal recessive and non-lethal autosomal dominant), the most common phenotype encountered by the anesthesia provider is the non-lethal autosomal dominant variation [3]. Even though the autosomal dominant variation of MD is associated with better clinical outcomes [5], there remains a significant anesthetic risk. Common MD

M. McCoy · M. Brown
Department of Anesthesiology and Perioperative Medicine, Nemours Children's Hospital, Wilmington, Delaware, USA
e-mail: Melinda.Brown@nemours.org; Matthew.mccoy@nemours.org

L. W. Averill
Radiology and Pediatrics, Sidney Kimmel Medical College, Thomas Jefferson University, Philadelphia, Pennsylvania, USA

Department of Radiology, Nemours Children's Hospital, Wilmington, Delaware, USA
e-mail: Lauren.averill@nemours.org

M. C. Theroux (✉)
Department of Anesthesiology and Perioperative Medicine, Nemours Children's Hospital, Wilmington, Delaware, USA

Anesthesiology and Pediatrics, Sidney Kimmel Medical College, Thomas Jefferson University, Philadelphia, Pennsylvania, USA
e-mail: mtheroux@nemours.org

M. C. Theroux (ed.), *Anesthesia Care of Patients with Skeletal Dysplasia*,
https://doi.org/10.1007/978-3-031-92501-6_11

characteristics that are associated with a difficult airway and increase the anesthetic risk include tracheomalacia, calcification of both the cricoid cartilage and the hyoid bone, short neck, fused cervical spine, and atlantoaxial instability [3, 6].

11.2 Genetics

Metatropic dysplasia is caused by a mutation in *TRPV4*, a calcium permeable non-selective cation channel encoded by chromosome 12, which is involved in numerous physiological processes and can lead to several musculoskeletal disorders [2, 7–9]. Although the exact function of *TRPV4* in the skeleton is unknown, activation has been shown to promote chondrogenesis, which may be a mechanism that leads to the development of skeletal dysplasia [2]. There are three forms of MD described: an autosomal-recessive form, both lethal and non-lethal, and a less severe, non-lethal autosomal dominant form [5, 10], indicating a span of severity. *TRPV4* mutations have been exhibited in both lethal and non-lethal MD [7]. Currently, more than 30 different *TRPV4* mutations have been described as causative for MD and other skeletal dysplasias, such as spondylometaphysealdysplasia and brachyolmia [9], as well as other neuromuscular conditions such as spinal muscular atrophy and Charocot-Marie-tooth [2, 7, 8].

11.3 Phenotype

Metatropic means "changing form," and MD is a changing dysplasia [8] due to the phenotypical characteristics that change from that of a short-limb/long-trunk dwarfism in infancy to a short-limb/short-trunk dwarfism [2] in later childhood as spinal changes progress [7]. Most patients with MD are born pre-term weighing less than 2 kg with long fingers and toes [11]. Common irregularities are observed in both the metaphyseal portion of the long bones and the vertebral bodies (platyspondyly) [7, 8]. Further clinical features include craniofacial abnormalities such as flat facies with mid-face hypoplasia and prominent forehead [8]. A short neck, tracheomalacia, deviated epiglottis and tracheal cartilage, torticollis, and cervical spine instability are additional features of MD [3]. The presence of a coccygeal tail has also been reported in some individuals [5, 8].

11.4 Radiologic Findings

Classic diagnostic radiologic findings include dumbbell shaped diaphysis, a halberd-shaped (battle ax with concave or convex blade) [2, 12] pelvis, flattening of the vertebral bodies (platyspondyly) and spinal deformity (often severe) [3, 5, 7, 13]. Severe kyphoscoliosis, odontoid hypoplasia, and cervical instability are very commonly seen in MD [14]. Other radiologic findings include wafer-thin vertebral bodies [8] with bulky pedicles [11], narrow sacrosciatic notches, short cupped ribs, and delayed ossification of the ischium and pubic bones [5]. In addition, early calcification of the hyoid and cricoid cartilage, epiphyseal ossification of the knees, delayed ossification and hypoplasia of C1, and irregularly contoured calcaneal bones may all be seen on radiographic analysis [2, 5].

11.5 Co-morbid Conditions

Joint limitations are frequently noted, as well as a narrowed thorax [5]. A triangular shaped, stiff chest wall leads to severe and frequent respiratory tract complications [13]. A major cause of mortality is respiratory distress related to severe kyphoscoliosis and a narrowed thorax [5, 12]. Osteopenia leading to fractures is also frequently encountered [9]. Medullary compression related to odontoid hypoplasia is of significant concern [15], and paraparesis associated with cervical instability and stenosis is possible [13].

11.6 Anesthesia Care

Airway management for patients with MD is of upmost importance as the airway becomes increasingly more difficult with age [3]. In one study, the majority of reported surgical procedures (48%) were related to either the cervical, lumbar, or thoracic spine; 13% were extremity procedures; 15% were medical imaging; 10% were general surgery procedures; 8% were otolaryngology (ENT) procedures; and 6% were miscellaneous [3] (Fig. 11.1). The same study reported that 68% of the intubations in 23 patients (who underwent a total of 188 anesthetics) used a video laryngoscope, 20% used a conventional laryngoscope (Miller or Mac blades), and 12% required a flexible bronchoscope (Fig. 11.2).

This means most patients with MD have a difficult airway. Calcification of both the hyoid and cricoid cartilage, along with cervical spine fusion (spontaneous fusion often superimposed on surgical decompression and fusion) and severe torticollis, can lead to head and neck rigidity, lending varying amounts of difficulty to both head position and mask ventilation [3]. In addition, patients with MD often present with their chin positioned in close proximity to their sternum (Fig. 11.3a), which does not facilitate easy mask ventilation [3]. Positioning the head and neck so that the mastoid process is aligned with the clavicle (Fig. 11.3b) is essential for two reasons: (1) it provides optimal alignment of the cervical spine; (2) it facilitates the chin pointing up, opening up the head and neck, favoring successful airway management [3]. It is important to know that surgical airway management of the patient with MD may be very difficult and require sternotomy, as the cricoid ring is often located below the sternal notch [3]. Severe respiratory tract compromise is often associated with the MD population related to multiple etiologies such as upper airway obstruction, spinal deformity, stiff chest wall, and bronchial and tracheal atresia [13]. Pulmonary functioning testing and sleep studies are warranted prior to procedures requiring anesthesia. Careful consideration should be given for

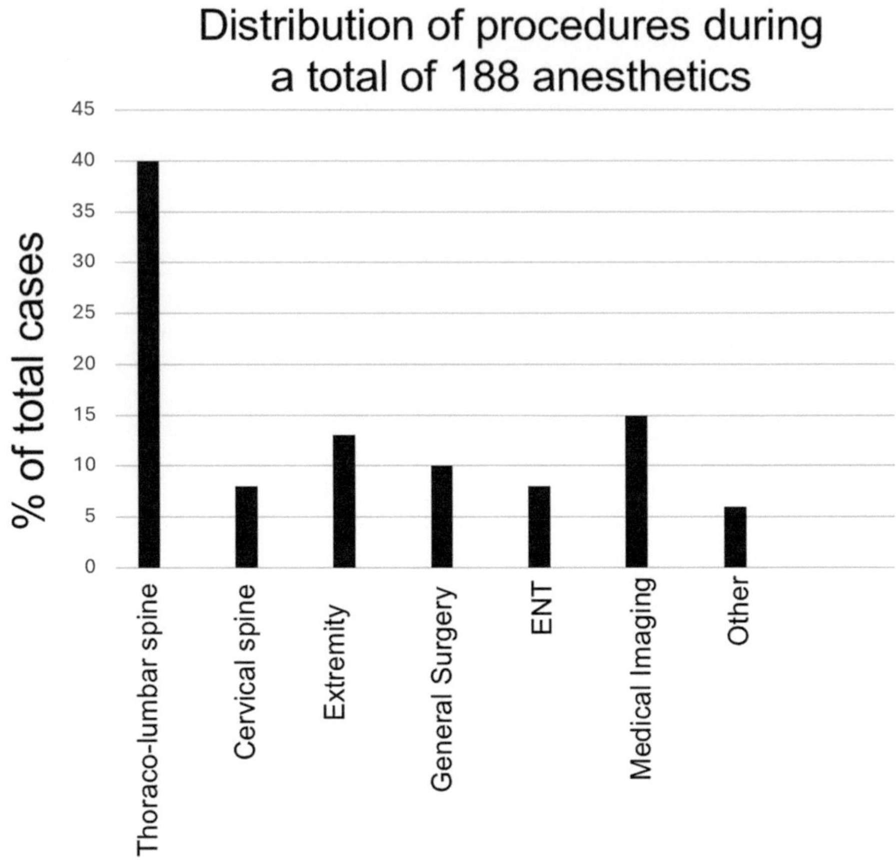

Fig. 11.1 Types of procedures in 23 patients [3]. About 61% of 188 total procedures were orthopedic (spine fusion and extremity surgery together). Average time for surgery was 112–124 min. The median time was 92 min (range, 15 min [wisdom teeth removal] to 528 min [anterior decompression and fusion of thoracolumbar spine via thoraco-abdominal approach]) [3]

postoperative respiratory support, such as bilateral positive airway pressure [13], and to the possible need for prolonged postoperative mechanical ventilation.

Unstable cervical spine is commonly seen in the MD population, and care must be taken to maintain the alignment of the mastoid process with the clavicle,[1] minimize head and neck movement [3]. Evoked potential monitoring during intubation should be considered [3]. This is especially important in offsite anesthetizing locations such as MRI. Roughly 15% of the anesthetics in patients with MD are to obtain medical imaging studies. Most, if not all, patients with MD will present to MRI for evaluation of odontoid hypoplasia, cervical instability, and cervical stenosis [13] with or without associated clinical signs of cord compression on flexion and

[1] Please refer to Chap. 16: Anesthetic Management of Spine Fusion.

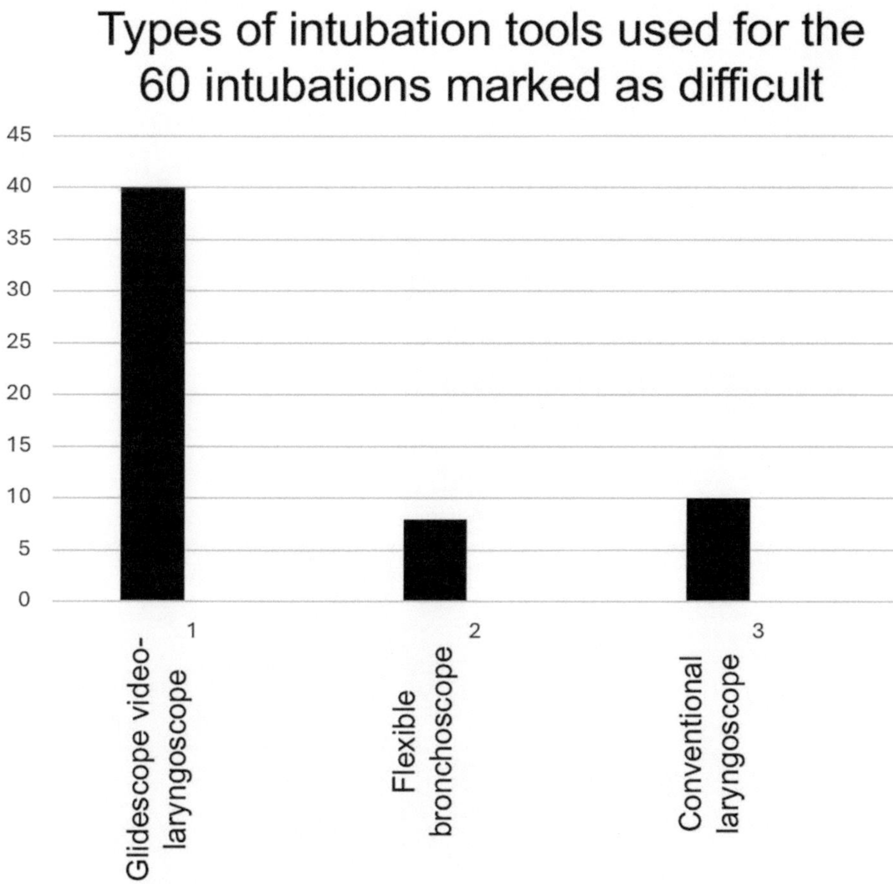

Fig. 11.2 Frequency of use of difficult airway tools for the 60 difficult intubations [3]. Glidescope was by far the most preferred tool as it allowed effective displacement of oropharyngeal structures. When Miller or Macintosh blades were used with success, the intubations were noted to be more difficult and required multiple attempts. Available information on the use of Miller or Macintosh blades includes one patient who had three anesthetics requiring tracheal intubation where each intubation took three attempts. In two other patients, poor visualization and use of significant cricoid pressure were noted; while, for another two patients, use of tongue retraction was mentioned but the number of attempts was not mentioned [3]

extension cervical spine films [12]. Unfortunately, evoked potential monitoring is difficult to impossible in this environment, so patient positioning and length of procedure must be optimized to decrease harm. In high-risk cases or "spine-at-risk" situations, maintaining mean arterial pressures at baseline or higher and abbreviating the total time of the imaging studies to an hour or less should be considered. During the flexion and extension studies, the head and neck must only be positioned by the anesthesia professional and should only be performed following the radiologist's approval of the scouting films.

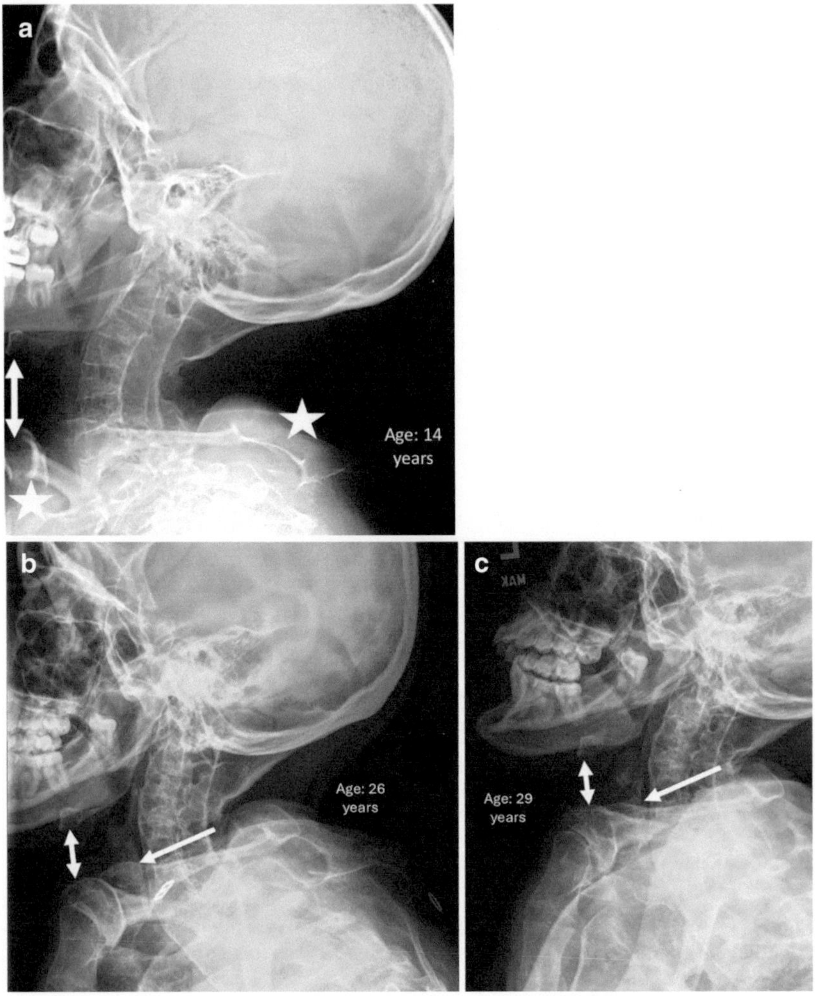

Fig. 11.3 Young woman with metatropic dysplasia who had progressively more difficult airway management during anesthesia. At 14 years old, a lateral X-ray of the cervical spine (**a**) shows complete bony fusion of the occiput and spine. Because of torticollis (not shown), her shoulders are high riding and obliquely imaged (stars). She has a short neck, with short distance between her hyoid bone and clavicles (double arrow). The imaging study was obtained at the same time as she experienced a difficult anesthetic, during which the procedure (dental) had to be canceled due to inability to intubate trachea. She was admitted to ICU for recovery. From age 14, the patient had been intubated successfully using video laryngoscope, but at age 26, she had become much more difficult an intubation using video laryngoscope. Cervical spine X-rays at 26 (**b**) and 29 years old (**c**) show worsening of short neck and high-riding shoulders with shorter distance between hyoid and clavicle (double arrows). Her laryngeal airway and trachea are not well seen. The cricoid cartilage is calcified (arrows), positioned at the level of the sternal notch. Currently, her intubations are done with her awake using optimal topicalization and sedation as tolerated. A sagittal reconstruction from CT of the chest (**d**) at 30 years old shows complete bony fusion of the spine and severe kyphosis centered in the lower thoracic spine. The calcified cricoid cartilage (arrow) is shown again at the level of the sternal notch (chevron), just above the manubrium (M)

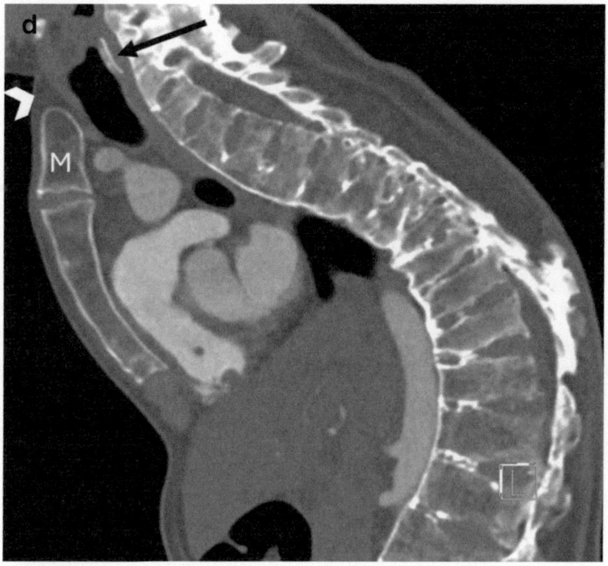

Fig. 11.3 (continued)

11.7 Neuraxial and Peripheral Nerve Blocks

Spinal and epidural anesthesia should be avoided in patients with MD as spinal canal stenosis is a common finding in this population [3]. Peripheral nerve blocks are gaining increasing popularity in the anesthetic management of complex medical patients, and plans for peripheral nerve blocks should include a discussion with the surgical team [14]. The need for neuromonitoring may dictate that the blocks should be performed at conclusion of the surgical procedure.

11.8 Case Illustration

A 31-year-old female diagnosed with metatropic dysplasia, who is a fully functioning individual with a full-time job as a teacher, has been under our care since age 3 years. Despite her active lifestyle and a full-time job, MD has significantly impacted her musculoskeletal system from an early age. Diagnosed in childhood, she has undergone numerous surgical procedures, with one of the earliest being a cervical spine fusion with halo placement at age 3 years. The majority of her surgeries were primarily orthopedic in nature and are too numerous to list. They have included multiple procedures on her spine and lower extremities, including her hips, many of which have been repeated as her dysplasia progressed. More recently, she has experienced renal stones, necessitating cystoscopy and lithotripsy.

11.8.1 Anesthesia Care

Up until the age of 12, her airway management during anesthesia was straightforward, with no recorded difficulties. An anesthetic record from that time notes successful intubation of the trachea using a curved Macintosh blade and conventional laryngoscopy. However, at age 15, during a general anesthetic for a dental procedure, the anesthesiologist encountered significant challenges. Despite multiple attempts with both conventional laryngoscopy and a video laryngoscope (GlideScope; Verathon, Bothell, WA, USA), intubation was unsuccessful. After several failed attempts accompanied by the onset of bloody secretions, the decision was made to awaken the patient and cancel the procedure. Face mask ventilation was adequate using a two-person technique, and the patient's recovery in the pediatric intensive care unit was uneventful.

Five years later, during her next anesthetic, a flexible bronchoscope was used successfully for intubation after induction of anesthesia. However, in a subsequent procedure 1 year later, intubation again was performed using the GlideScope, this time with the added assistance of tongue retraction with a piece of gauze. Following this, her anesthetic care continued with similar intubation techniques, always involving the same anesthesiologist and a nurse anesthetist familiar with the patient. During this period, the patient began to request this specific anesthesiologist, even to the extent of postponing surgery if the anesthesiologist was unavailable.

At age 26, during another orthopedic procedure, the previously successful GlideScope technique failed to secure the airway on the first attempt, and mask ventilation proved more difficult than for prior anesthetics. Fortunately, a second intubation attempt was successful with the tongue extracted more than during the first attempt. But this incident led to a change in her anesthetic management strategy; the anesthesiologist formulated a plan for awake intubation with meticulous topicalization and sedation. Since then, she has undergone five additional general anesthetics for combined orthopedic and urological procedures, all of which were managed successfully using this technique.

Despite the challenges posed by her skeletal dysplasia and associated disabilities, the patient continues to thrive in her professional life, maintaining a full-time teaching position. Her case exemplifies the importance of tailored, adaptive anesthetic strategies for individuals with complex airway anatomy due to skeletal dysplasias. Her cervical spine radiographs at different stages of difficulty with intubations are below (Fig. 11.3).

References

1. Maroteaux P, Spranger J, Wiedemann HR. Metatrophic dwarfism. Arch Kinderheilkd. 1966;173:211–26.
2. Nemec SF, Cohn DH, Krakow D, Funari VA, Rimoin DL, Lachman RS. The importance of conventional radiography in the mutational analysis of skeletal dysplasias (the TRPV4 mutational family). Pediatr Radiol. 2012;42:15–23.

3. Theroux MC, Lopez M, Olszewsky PJ, DiCindio S, Arai L, Ditro C, et al. Metatropic dysplasia-a skeletal dysplasia with challenging airway and other anesthetic concerns. Paediatr Anaesth. 2017;27:596–603.
4. Beck M, Roubicek M, Rogers JG, Naumoff P, Spranger J. Heterogeneity of metatropic dysplasia. Eur J Pediatr. 1983;140:231–7.
5. Geneviève D, Le Merrer M, Feingold J, Munnich A, Maroteaux P, Cormier-Daire V. Revisiting metatropic dysplasia: presentation of a series of 19 novel patients and review of the literature. Am J Med Genet A. 2008;146A:992–6.
6. Sitoula P, Mackenzie WG, Shah SA, Thacker M, Ditro C, Holmes L Jr, et al. Occipitocervical fusion in skeletal dysplasia: a new surgical technique. Spine (Phila Pa. 1976). 2014;39:E912–8.
7. Andreucci E, Aftimos S, Alcausin M, Haan E, Hunter W, Kannu P, et al. TRPV4 related skeletal dysplasias: a phenotypic spectrum highlighted byclinical, radiographic, and molecular studies in 21 new families. Orphanet J Rare Dis. 2011;6:37.
8. Narayanan DL, Bhavani GS, Girisha KM, Phadke SR. Metatropic dysplasia with a novel mutation in TRPV4. Indian Pediatr. 2016;53:735–7.
9. Tomatsu S, Averill LW, Sawamoto K, Mackenzie WG, Bober MB, Pizarro C, et al. Obstructive airway in morquio A syndrome, the past, the present and the future. Mol Genet Metab. 2016;117:150–6.
10. Camacho N, Krakow D, Johnykutty S, Katzman PJ, Pepkowitz S, Vriens J, et al. Dominant TRPV4 mutations in nonlethal and lethal metatropic dysplasia. Am J Med Genet A. 2010;152A:1169–77.
11. Hall CM, Elçioglu NH. Metatropic dysplasia lethal variants. Pediatr Radiol. 2004;34:66–74.
12. McCray BA, Schindler A, Hoover-Fong JE, et al. Autosomal dominant TRPV4 disorders. 2014 May 15 [Updated 2020 Sep 17]. In: Adam MP, Feldman J, Mirzaa GM, et al., editors. GeneReviews® [Internet]. Seattle: University of Washington, Seattle; 1993–2024. https://www.ncbi.nlm.nih.gov/books/NBK201366/.
13. Leet AI, Sampath JS, Scott CI Jr, MacKenzie WG. Cervical spinal stenosis in metatropic dysplasia. J Pediatr Orthop. 2006;26:347–52.
14. Eiszner JR, Atanda A Jr, Rangavajjula A, Theroux M. A case series of peripheral nerve blocks in pediatrics and young adults with skeletal dysplasia. Paediatr Anaesth. 2016;26:553–6.
15. Shohat M, Lachman R, Rimoin DL. Odontoid hypoplasia with vertebral cervical subluxation and ventriculomegaly in metatropic dysplasia. J Pediatr. 1989;114:239–43.

Campomelic Dysplasia

12

Divya Dixit, Lauren W. Averill, and Mary C. Theroux

12.1 Introduction

Campomelic dysplasia is a rare form of short-limb dwarfism. This skeletal dysplasia is characterized by short stature, cartilaginous weakness of the airway, bowing of long bones, kyphoscoliosis, cervical spine instability, typical facial features, and early death [1]. Life-sustaining interventions, such as tracheostomy, may be needed to prolong survival in severe presentations. In this chapter, we describe the multisystem involvement in campomelic dysplasia and anesthetic considerations and management of children with this form of dwarfism.

D. Dixit (✉) · M. C. Theroux
Department of Anesthesiology and Perioperative Medicine, Nemours Children's Hospital, Wilmington, Delaware, USA

Anesthesiology and Pediatrics, Sidney Kimmel Medical College, Thomas Jefferson University, Philadelphia, Pennsylvania, USA
e-mail: Divya.Dixit@nemours.org; mtheroux@nemours.org

L. W. Averill
Department of Radiology and Pediatrics, Sidney Kimmel Medical College, Thomas Jefferson University, Philadelphia, Pennsylvania, USA

Department of Radiology, Nemours Children's Hospital, Wilmington, Delaware, USA
e-mail: Lauren.averill@nemours.org

M. C. Theroux (ed.), *Anesthesia Care of Patients with Skeletal Dysplasia*,
https://doi.org/10.1007/978-3-031-92501-6_12

12.2 Genes and Transmission

Campomelic dwarfism was first described by Dr. Pierre Maroteaux in 1971 [2]. The term "campomelic," or "camptomelic," is of Greek origin, and it means "bent limb." The incidence of campomelic dysplasia is reported to be 1:40,000 to 1:80,000. It may be inherited in an autosomal dominant manner, although most cases have a de novo mutation. Campomelic dysplasia is caused by mutation or chromosomal rearrangements affecting the sex-determining region Y gene (SRY)-box transcription factor 9 (*SOX9*) gene, which is located on chromosome 17q [3, 4]. *SOX9*, a member of the SOX family of transcription factors, regulates chondrocyte differentiation, regulates the cartilage development pathway, and is important for neural crest cell derived endochondral bone formation. It plays a role in both bone formation and testicular development.

12.3 Phenotypic Features

Distinctive facial anomalies found in children with campomelic dysplasia include a macrocephalic dolichocephalic head, prominent forehead, midface hypoplasia, flattened supraorbital ridges, short palpebral fissures, overturned ala nasi, low-set ears, and micrognathia. Also, depressed nasal bridge, hypertelorism, and long philtrum are commonly present facial features [5]. Cleft palate is present in two-thirds of these children [1, 2]. Pierre Robin sequence (micrognathia, cleft palate, and glossoptosis) is a common association [2, 6]. Generally, patients with campomelic dysplasia have normal intelligence with mild-to-moderate learning difficulties, and global developmental delays have been reported.

12.4 Musculoskeletal Features

Campomelic dysplasia is characterized by bowing and shortening of the long bones. Frequently, the lower extremity bones, femur, tibia, and fibula are bowed symmetrically [2, 7]. Bowing of the femur at the proximal one-third of the femoral shaft and bowing of the tibia at the distal two-third of the tibial shaft are prominent characteristics. Delayed ossification of distal femoral and proximal tibial epiphyses is a frequent finding. Some individuals may have absence of any appreciable bending of the long bones while having other major features of campomelic dysplasia. The obligatory feature of bowed long bones is absent in about 10% of individuals with campomelic dwarfism, and such individuals are referred to as having "acampomelic" campomelic dysplasia [8, 9]. Dimpling of skin over the apex of bent tibial curve is a characteristic clinical indicator of campomelic dysplasia. These children have short stature with short limbs at birth; however, with increasing kyphoscoliosis, the trunk becomes progressively shorter relative to the limbs.

Other common features of campomelic dysplasia are 11 pairs of ribs and hypoplastic scapulae. Abnormal cervical vertebrae are present in 77% of patients with campomelic dysplasia, including hypoplastic vertebral bodies, non-mineralized thoracic pedicles, and cervical thoracic kyphoscoliosis (Fig. 12.1) [1, 10]. Cervical spine kyphosis is common and related to failure of formation of the anterior cervical vertebral bodies. It can lead to spinal cord compression, which may need surgical fixation with instrumentation. Affecting a significant number of patients with campomelic dysplasia, thoracic kyphoscoliosis is progressive and may require various surgical procedures, including traction halo placement, anterior thoracoscopic release, and/or posterior spine fusion (Fig. 12.2) [11, 12]. In a series of eight patients with campomelic dysplasia, thoracic hyperkyphosis averaging 126° was found in six (75%) of the patients, while cervical kyphosis averaging 66° was reported in three (38%) [10]. Vertically narrow pelvic iliac bones may be associated in these patients. Non-ossification of inferior pubic rami and short ischia are commonly present. Hip dislocations and bilateral talipes equinovarus may be associated with

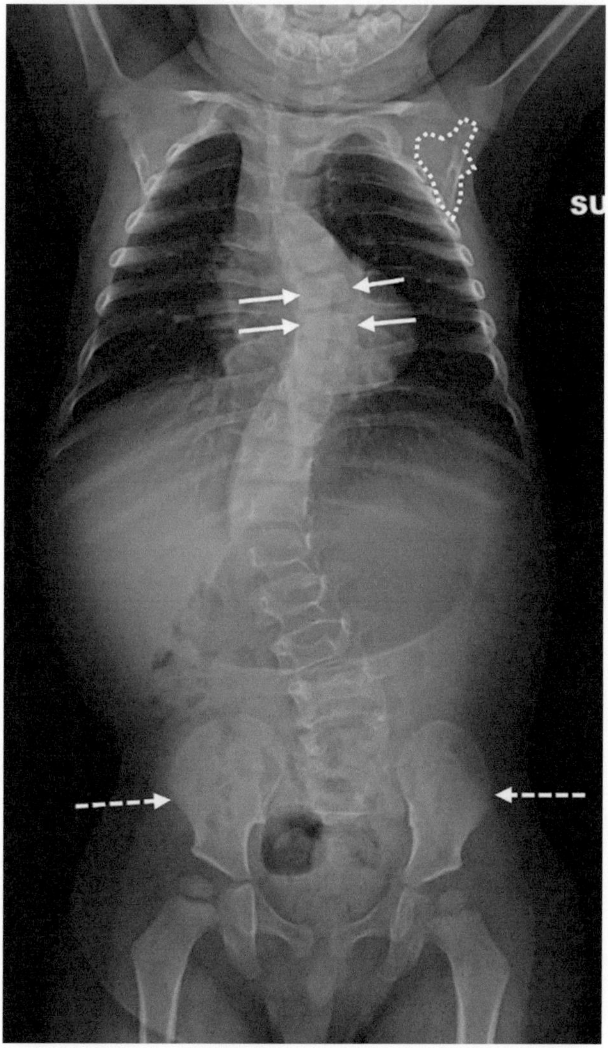

Fig. 12.1 X-ray image of supine anteroposterior view of the spine and pelvis of a 20-month-old child with campomelic dysplasia showing thoracic kyphoscoliosis, hypoplastic thoracic vertebral pedicle (solid arrows), scapular hypoplasia (dotted outline), bell-shaped chest, and narrow iliac wings (dashed arrows)

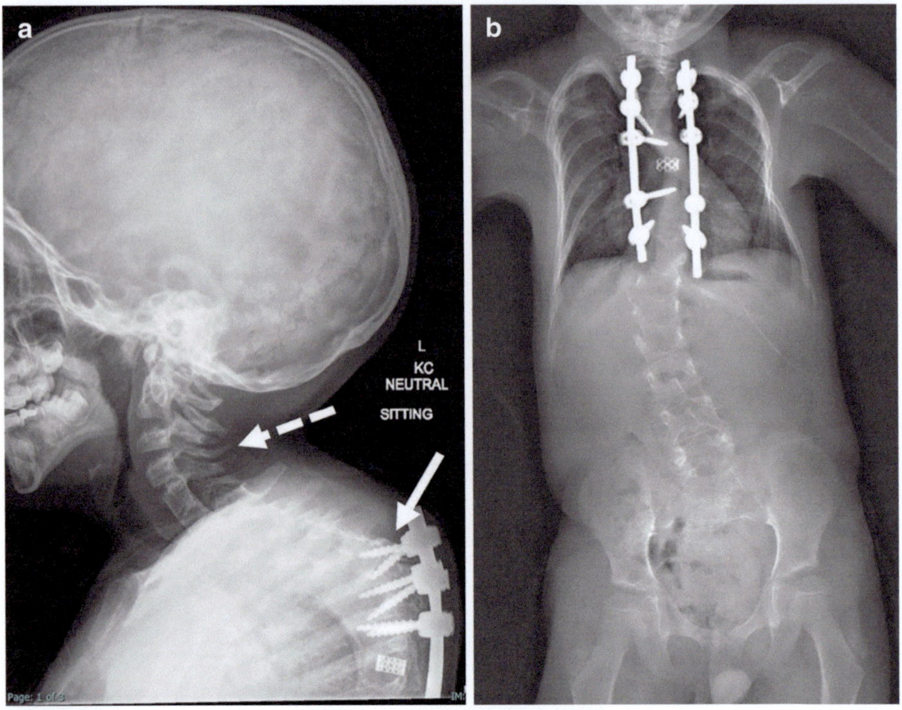

Fig. 12.2 Spine lateral (**a**) and frontal (**b**) x-ray images of an 8-year-old child with campomelic dysplasia showing severely kyphotic thoracic spine (solid arrow) surgically fused with bilateral rods and pedicle screws (**b**). Prominent cervical lordosis is present in neutral position (**a**) (dashed arrow)

Table 12.1 Common radiological findings in patients with campomelic dysplasia [1, 5]

Long bones	Bowing of femur and tibia, hypoplastic fibula
	Delayed ossification of distal femoral and proximal tibial epiphyses
Scapula	Hypoplastic scapula
Cervical vertebra	Hypoplastic cervical vertebrae, cervical kyphosis
Thoracic vertebra	Non-mineralized thoracic vertebral pedicles, thoracic kyphosis
Thorax	Small, bell-shaped chest
	11 pairs of ribs, slender ribs
Pelvis	Vertical narrow iliac bones, dislocated hips, defective ischiopubic ossification, hypoplastic pubic bones
Feet	Bilateral club feet, short phalanges
Hands	Short first metacarpals, short phalanges

campomelic dysplasia [13, 14]. The skeletal disorders mentioned above frequently require surgical correction, often involving multiple exposures to anesthesia.

Radiological findings suggestive of campomelic skeletal dysplasia diagnosis are summarized in Table 12.1.

12.5 Airway, Tracheobronchial Tree, and Pulmonary System

Historically, most newborns with campomelic dysplasia die of respiratory insufficiency in the neonatal period. In Mansour et al.'s 1995 study of 36 patients, there were three survivors; 77% of deaths were caused by severe respiratory distress in the neonatal period with 90% of the patients dying before age 2 years [1]. The most characteristic abnormality in patients with campomelic dysplasia is the lack of development of the cartilage rings supporting the tracheobronchial tree that normally keeps the respiratory passages open. Deficient cartilage development (reduced number, fused and fenestrated anteriorly, incomplete posteriorly, absence in smaller bronchi) may cause the collapse of the soft tracheobronchial passages, producing extremely narrow airways and respiratory insufficiency [2]. Pulmonary hypoplasia, narrow and easily collapsible airways, and mucus occlusion often lead to aspiration, atelectasis, apneic spells, pneumonia, and death in the neonatal period. Tracheobronchomalacia can cause obstruction both during the inspiratory and expiratory phase of respiration. Signs and symptoms suggestive of tracheomalacia and tracheobronchomalacia in children include noisy breathing, recurrent respiratory infections, cough, barky cough, difficulty feeding, aspiration, respiratory distress, oxygen dependence, and life-threatening events [15]. This is significant for the anesthesiologist, as it may present as chronic airway obstruction preoperatively and manifest as subcostal and chest wall retractions postoperatively. At birth, the child may require a tracheostomy and long-term ventilation for the first few years of life. Houston et al. and Mansour et al. (2002) have described one patient in each of their cohorts needing tracheostomy out of 17 and 5 patients, respectively, with campomelic dysplasia [2, 5]. There are risks associated with tracheostomy, particularly in small children, such as changes in size and length of tracheostomy tube, tracheal injury, tracheal stenosis, fistula formation, granulation tissue, recurrent infection, and accidental decannulation [16]. However, not all patients with campomelic dysplasia will require a tracheostomy. Continuous positive airway pressure could provide effective treatment for moderate-to-severe tracheobronchomalacia by maintaining above-ambient atmospheric pressure in the tracheal lumen, preventing airway collapse during expiration and decreasing resistance to airflow during expiration when malacic airway segments typically collapse [17]. Progressive cervicothoracic kyphosis and spinal cord compression present in this syndrome may further depress respiration. The lungs and chest may be small, narrow, and bell-shaped, and restrictive lung disease may be present.

12.6 Other Co-morbid Conditions

In addition to laryngomalacia and tracheobronchomalacia, Pierre Robin sequence may be present in campomelic dysplasia. Individuals with campomelic dysplasia may also have mixed hearing loss. High incidence of otitis media, poor development of auditory ossicles, and atypical skull shape are contributory factors to conductive hearing loss, whereas sensorineural hearing loss would be related to cochlear malformations [18]. Any suspicion of hearing loss or recurrent ear infections should prompt referral to an ear, nose, and throat surgeon and audiologist for further investigation and management. In some cases, hearing aids or cochlear implants may be appropriate. Pierre Robin sequence may require treatment with mandibular distraction and cleft palate repair [19, 20].

Congenital heart defects occur in 25% of patients with campomelic dysplasia [2]. Common anomalies include patent ductus arteriosus, patent foramen ovale, atrial and ventricular septal defects, and tetralogy of Fallot. Genitourinary abnormalities such as hydronephrosis and bilateral hydroureter are present in one-third of the patients. Hypoplastic cystic kidney, renal hypoplasia, ureteral stenosis, and renal calculi also have been described in the literature. Lack of a functioning *SOX9* gene may lead to arrested development of male genitalia, testicular development, and male hormones. Therefore, 75% of genotypic males with XY karyotype have female or ambiguous genitalia [1]. Generalized hypotonia has been noted in children with campomelic dysplasia, and developmental milestones may be delayed. Central nervous system abnormalities include hydrocephalus, macrocephaly, and polygyria; absence of olfactory tract or bulbs was reported in about half of autopsied cases, often with dilatation of lateral ventricles [2].

12.7 Anesthesia Care

In infancy and childhood, patients would likely have undergone medical evaluation for upper airway obstruction and collapsibility of the tracheobronchial tree including radiological studies and airway fluoroscopy (Fig. 12.3). Rigid or flexible bronchoscopy provides accurate diagnosis for tracheobronchomalacia and may require general anesthesia. Dynamic 3-phase bronchoscopy is considered the gold standard for diagnosis of tracheobronchomalacia [16]. Greater than 50% narrowing of the dynamic airway lumen during forced exhalation or coughing is considered diagnostic of tracheobronchomalacia by most clinicians [15]. Greater than 75% of airway lumen obstruction during exhalation is present in most symptomatic children with tracheomalacia. Airway management, especially during induction and emergence of anesthesia, is critical to prevent complete airway collapse. Graded inhalation induction with incremental doses of intravenous medications such as dexmedetomidine, ketamine, or remifentanil with assisted positive pressure ventilation provides a balanced approach to maintain the airway tone [21]. Upper airway collapsibility may worsen with propofol because of depression of central respiratory output to upper airway pharyngeal dilator muscles and upper airway reflexes [22]. In addition, because Pierre Robin sequence is commonly associated, preparation should be made for a potential difficult airway. Need for palatoplasty surgery is uncommon as the survival rate is poor beyond infancy [19]. However, as the number of survivors is increasing with improved diagnosis and perinatal care, these procedures are more likely to occur. Timing of the palatoplasty requires consideration;

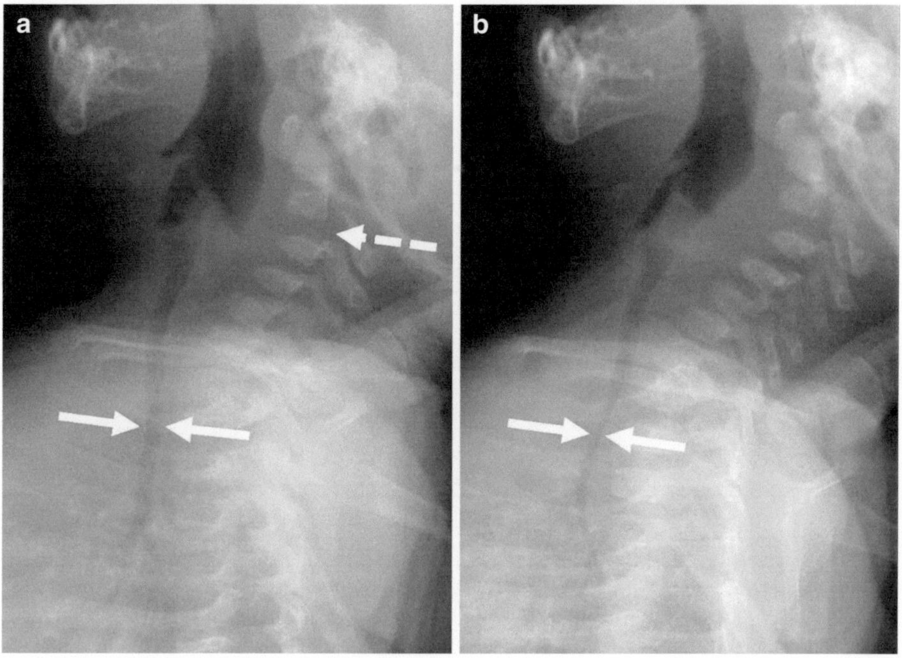

Fig. 12.3 Two selected images from dynamic airway fluoroscopy of a 4-month-old child with campomelic dysplasia in respiratory distress, showing the anteroposterior diameter of the trachea in lateral projection of approximately 3.4 mm (**a**, solid arrows) collapsing to about 1 mm distally (**b**, solid arrows), consistent with severe tracheobronchomalacia. Kyphosis of cervical spine is shown here as well (dashed arrow)

respiratory status of the patient should be stable prior to proceeding with palatoplasty. Early tracheostomy followed by mandibular distraction and cleft palate repair may be a sequence of procedures to consider, as was recommended in a limited, 2-case study of campomelic dysplasia [20].

Careful clinical examination and radiological evaluation is necessary prior to any surgery to rule out cervical spine instability and cord compression (Fig. 12.4). This may be done by evaluating neck radiographs in flexion and extension views as well as with further imaging studies such as magnetic resonance imaging, if necessary. It may be important to maintain the patient's head and neck in neutral position on the operating room table for induction of anesthesia and tracheal intubation [9]. Orthopedic and neurosurgical input regarding cervical spine stability is invaluable in planning for any surgical procedure. If spinal cord compression is suspected, surgical treatment to relieve cord compression should be prioritized (as recommended by neurosurgery and orthopedics). Such procedures may include cervical laminectomy with decompression of the spinal cord, halo traction placement, or cervical spine fusion for stabilization. Neuromonitoring during such

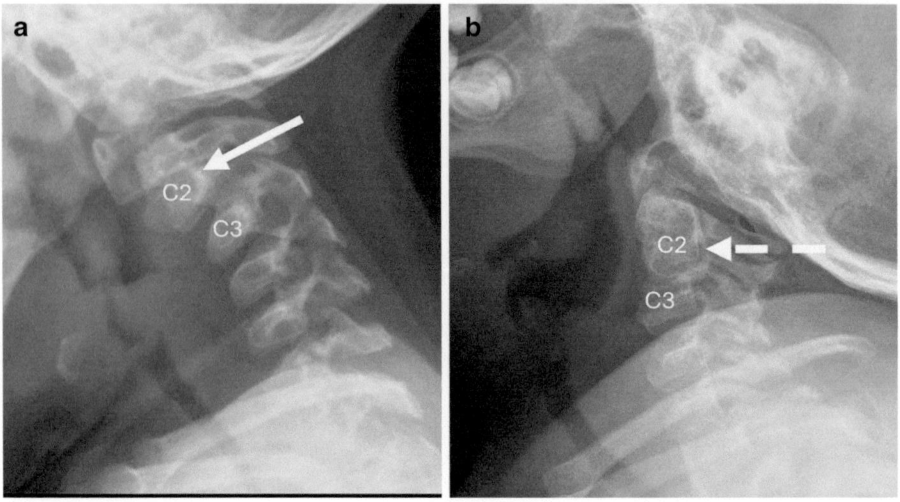

Fig. 12.4 X-ray images of the lateral cervical spine in both flexion (**a**) and extension (**b**) of a 3-year-old child with campomelic dysplasia showing anterior subluxation at C2–3 measuring about 4.4 mm (solid arrow). Lateral view in extension shows retro subluxation at C2–3 measuring about 4.0 mm (dashed arrow)

procedures mandates a total intravenous anesthesia (even though some institutions may have a modified approach) and is considered the standard practice by most institutions. Both motor and sensory evoked potentials should be monitored to assess spinal cord integrity.

Patients may be placed in a halo traction device, and subsequent anesthetics will require anesthetizing the child with the halo device in place (Fig. 12.5). It poses a greater challenge to the anesthesiologist and difficulty with inhalation induction, mask ventilation and tracheal intubation should be anticipated. Alternative airway devices such as video laryngoscopes, flexible bronchoscope, and supraglottic airway devices should be readily available and used as necessary. Direct laryngoscopy has the potential to cause an injury to an already compromised spinal cord. At our institute, we frequently use video laryngoscope such as GlideScope (Verathon, Bothell, WA, USA) or C-MAC video laryngoscope (Karl Storz, Tuttlingen, Germany) to minimize the possibility of the cervical spinal cord compression with direct laryngoscopy for tracheal intubation [23, 24]. Other evaluations should include an echocardiogram to evaluate for possible congenital heart disease. Laboratory tests prior to anesthesia are considered based on the patient's health status and co-morbid conditions.

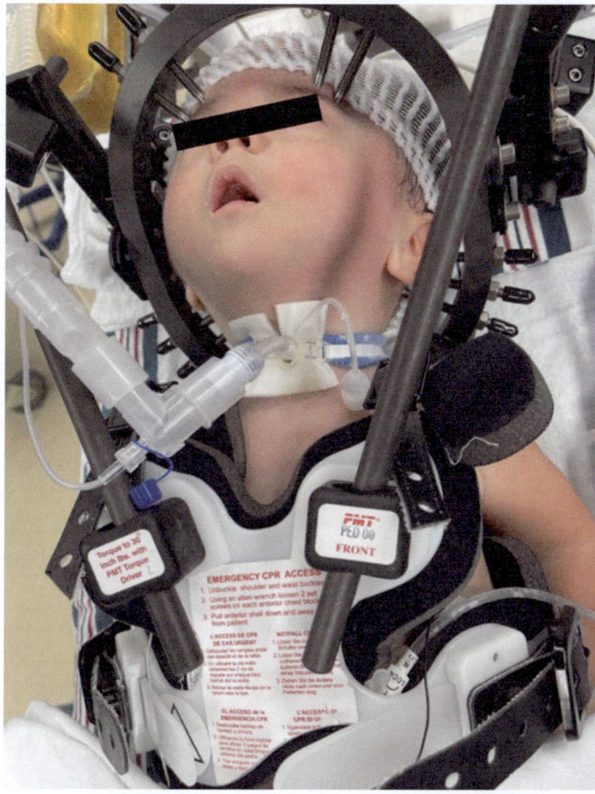

Fig. 12.5 Halo traction device in a 2-year-old child with campomelic dysplasia. Also seen is tracheostomy in situ

12.8 Postoperative Care

The patient's airway difficulties may dictate the recovery location after surgery and anesthesia. Noninvasive positive pressure ventilation may be required in the immediate postoperative period to stent the airways open and maintain air exchange while the child is recovering from anesthesia. However, observation in the intensive care unit (ICU) postoperatively may be appropriate for patients with severe dynamic airway obstruction preoperatively. Children with existing tracheostomy may recover on regular ward or floor specialized in caring for patients with tracheostomy and can provide supportive ventilation, as needed. Other factors that may require ICU care are the need for monitoring of neurological status following neurosurgical procedures such as cervical cord decompression. Such decisions are based on the level of care anticipated in the postoperative period balanced against the availability of skilled care in different patient areas of the hospital. Nelson et al. reported failed tracheal extubation requiring tracheostomy in two neonates with campomelic dysplasia and Pierre Robin sequence: one post mandibular distraction surgery, and another post tracheal intubation secondary to respiratory distress at birth. In both cases, failure to extubate the trachea was thought to be related to a combination of factors, including hypotonia, glossoptosis, severe bronchomalacia, and main stem bronchus dynamic collapse [20]. In both cases, the infants needed tracheostomy to manage the airway obstruction and to establish ventilation.

12.9 Case Illustration

A 14-month-old male weighing 8.5 kg with a diagnosis of *SOX9* haploinsufficiency with campomelic dysplasia presented for operative repair of a bilateral cleft palate. His birth weight was 8 pounds 12 ounces. He was born full-term but transferred shortly after birth to a neonatal ICU where he was admitted for about 4 days. He received supplemental oxygen but did not require tracheal intubation and then was discharged home. Regarding his growth and development, he did not walk or sit independently but scooted on his abdomen on the floor using his hands and feet to propel himself. He received physical therapy and occupational therapy. The child also had hearing loss.

His past medical history was significant for reactive airway disease, Pierre Robin sequence, tracheobronchomalacia, and clubfeet. There was no prior general anesthesia encounter, having had tendon lengthening under local anesthesia to treat clubfoot at an outside hospital. He had numerous musculoskeletal abnormalities, including macrocephaly, mild kyphosis at the C2–3 level, vertical defect in the posterior elements of C2 without instability, widening of the atlantodens interval in flexion, and scoliosis, which was being managed with the Wilmington jacket [25]. Cardiac examination revealed a normal cardiac segmental anatomy and function on echocardiogram, patent foramen ovale with left-to-right flow, and a structurally normal heart. He had difficulty swallowing with the history of choking. Barium swallow study showed trace laryngeal penetration with thin liquid mixed formula and occasional choking episodes while eating. Preoperatively, airway fluoroscopy findings were consistent with severe tracheobronchomalacia, and an awake transnasal flexible laryngoscopy revealed an omega-shaped epiglottis with mild arytenoid redundancy, prolapse, and laryngomalacia. On clinical examination, the child had features of Pierre Robin sequence, micrognathia, and cleft palate. He had significant pectus excavatum with chronic baseline noisy breathing; however, his chest was clear to auscultation with the relevant history of recovery from a recent upper respiratory infection.

For his cleft palate repair, anesthesia was initiated with an inhalational induction with oxygen and sevoflurane, and a peripheral intravenous access was established. The trachea was intubated with a 3.5-mm micro cuff oral RAE (Ring, Adair, and Elwyn) endotracheal tube using GlideScope. The endotracheal tube position was verified with flexible bronchoscope showing endotracheal tube tip well above the carina and marked compression of trachea in the anteroposterior dimension. Anesthesia was maintained with oxygen, air, and isoflurane. Seven milligram of rocuronium and 1.5 mcg/kg/h fentanyl were administered during the 5-h procedure.

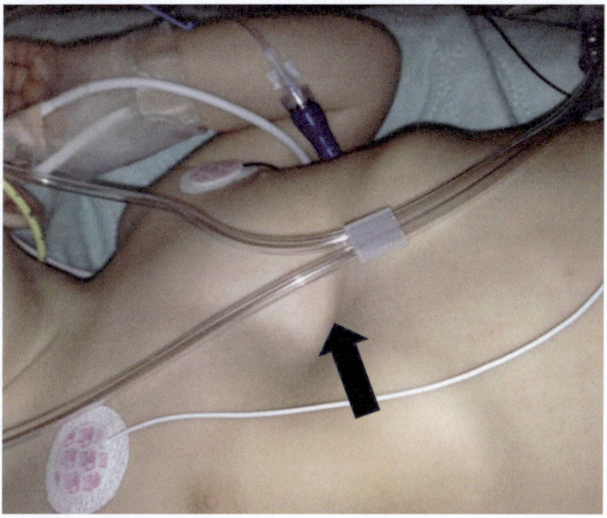

Fig. 12.6 Airway obstruction image of a child with campomelic dysplasia in respiratory distress. Significant sternal retractions (arrow) occurred following extubation of trachea as described in the case description

Intraoperatively, there were no complications noted, and extubation of the trachea at the end of the procedure was planned. Following reversal of neuromuscular blockade and establishment of regular spontaneous respirations, the patient's trachea was extubated when fully awake and crying. Upon planned arrival to the ICU, the patient started to show signs of severe respiratory distress. He developed significant upper airway obstruction with pronounced sternal retractions (Fig. 12.6). Initially, the patient was able to maintain oxygen saturation, but it became evident that he was not sustaining adequate ventilation and oxygenation despite receiving supplemental oxygen through a high flow nasal cannula (Vapotherm, Exeter, NH) at a flow rate of 12 L/min and FiO_2 of 100%. Additionally, dexamethasone and nebulized hypertonic saline were administered to treat post-surgical swelling. Ultimately, the patient required tracheal re-intubation to establish patent airway and ventilation. Persistent upper airway swelling and bronchospasm continued for the next several days. The patient received several more doses of dexamethasone for treatment of airway inflammation. The patient was persistently febrile with no obvious source of infection; however, antibiotics showed a decrease in fever curve. Trachea was extubated to heliox (mixture of helium and oxygen) and subsequently weaned to high flow nasal canula on postoperative day 5 with anesthesia stand-by. The patient was transferred to the floor on postoperative day 7 and discharged on postoperative day 11 following opioid and benzodiazepine wean.

This case illustrates the severity of respiratory complications arising in children with campomelic dysplasia from the underlying tracheobronchomalacia, resulting in severe airway obstruction and requiring tracheal re-intubation, prolonged mechanical ventilation in ICU, and an extended hospital stay.

References

1. Mansour S, Hall CM, Pembrey ME, Young ID. A clinical and genetic study of campomelic dysplasia. J Med Genet. 1995;32:415–20.
2. Houston CS, Opitz JM, Spranger JW, Macpherson RI, Reed MH, Gilbert EF, et al. The campomelic syndrome: review, report of 17 cases, and follow-up on the currently 17-year-old boy first reported by Maroteaux et al in 1971. Am J Med Genet. 1983;15:3–28.
3. Foster JW, Dominguez-Steglich MA, Guioli S, Kwok C, Weller PA, Stevanović M, et al. Campomelic dysplasia and autosomal sex reversal caused by mutations in an SRY-related gene. Nature. 1994;372:525–30.
4. Wagner T, Wirth J, Meyer J, Zabel B, Held M, Zimmer J, et al. Autosomal sex reversal and campomelic dysplasia are caused by mutations in and around the SRY-related gene SOX9. Cell. 1994;79:1111–20.
5. Mansour S, Offiah AC, McDowall S, Sim P, Tolmie J, Hall C. The phenotype of survivors of campomelic dysplasia. J Med Genet. 2002;39:597–602.
6. Unger S, Scherer G, Superti-Furga A. Campomelic dysplasia. In: Adam MP, Mirzaa GM, Pagon RA, Wallace SE, Bean LJH, Gripp KW, et al., editors. GeneReviews®. Seattle: University of Washington, Seattle; 1993–2024.
7. Jain V, Sen B. Campomelic dysplasia. J Pediatr Orthop B. 2014;23:485–8.
8. Gopakumar H, Superti-Furga A, Unger S, Scherer G, Rajiv PK, Nampoothiri S. Acampomelic form of campomelic dysplasia with SOX9 missense mutation. Indian J Pediatr. 2014;81:98–100.
9. Pasupathy M, Radhakrishnan V, Adenwalla HS, Narayanan PV. A case report of acampomelic campomelic dysplasia and operative difficulties in cleft palate reconstruction. Indian J Plast Surg. 2016;49:253–7.
10. Coscia MF, Bassett GS, Bowen JR, Ogilvie JW, Winter RB, Simonton SC. Spinal abnormalities in camptomelic dysplasia. J Pediatr Orthop. 1989;9:6–14.
11. Pourtaheri S, Shah SA, Ditro CP, Holmes L Jr, Mackenzie WG. Preoperative halo-gravity traction with and without thoracoscopic anterior release for skeletal dysplasia patients with severe kyphoscoliosis. J Child Orthop. 2016;10:135–42.
12. Thomas S, Winter RB, Lonstein JE. The treatment of progressive kyphoscoliosis in camptomelic dysplasia. Spine (Phila Pa 1976). 1997;22(12):1330–7.
13. Khoshhal K, Letts RM. Orthopaedic manifestations of campomelic dysplasia. Clin Orthop Relat Res. 2002;401:65–74.
14. Ray S, Bowen JR. Orthopaedic problems associated with survival in campomelic dysplasia. Clin Orthop Relat Res. 1984;185:77–82.
15. Fraga JC, Jennings RW, Kim PC. Pediatric tracheomalacia. Semin Pediatr Surg. 2016;25:156–64.
16. Kamran A, Jennings RW. Tracheomalacia and tracheobronchomalacia in pediatrics: an overview of evaluation, medical management, and surgical treatment. Front Pediatr. 2019;7:512.
17. Davis S, Jones M, Kisling J, Angelicchio C, Tepper RS. Effect of continuous positive airway pressure on forced expiratory flows in infants with tracheomalacia. Am J Respir Crit Care Med. 1998;158:148–52.
18. Takahashi H, Sando I, Masutani H. Temporal bone histopathological findings in campomelic dysplasia. J Laryngol Otol. 1992;106:361–5.
19. Narimatsu K, Iida A, Kobayashi T. Palatoplasty for the patient with campomelic dysplasia-report of a case and review of the literature. Cleft Palate Craniofac J. 2022;59:132–6.
20. Nelson ME, Griffin GR, Innis JW, Green GE. Campomelic dysplasia: airway management in two patients and an update on clinical-molecular correlations in the head and neck. Ann Otol Rhinol Laryngol. 2011;120:682–5.
21. Lalwani P, Gupta B. A rare and challenging case of anaesthetic management of paediatric laryngo-tracheobronchomalacia for supraglottoplasty. Int J Res Med Sci. 2023;11:3080–2.
22. Hillman DR, Walsh JH, Maddison KJ, et al. Evolution of changes in upper airway collapsibility during slow induction of anesthesia with propofol. Anesthesiology. 2009;111:63–71.

23. Suppan L, Tramèr MR, Niquille M, Grosgurin O, Marti C. Alternative intubation techniques vs Macintosh laryngoscopy in patients with cervical spine immobilization: systematic review and meta-analysis of randomized controlled trials. Br J Anaesth. 2016;116:27–36.
24. Singleton BN, Morris FK, Yet B, Buggy DJ, Perkins ZB. Effectiveness of intubation devices in patients with cervical spine immobilisation: a systematic review and network meta-analysis. Br J Anaesth. 2021;126:1055–66.
25. Bunnell WP, MacEwen GD, Jayakumar S. The use of plastic jackets in the non-operative treatment of idiopathic scoliosis. Preliminary report. J Bone Joint Surg Am. 1980;62:31–8.

Microcephalic Primordial Dwarfism 13

Divya Dixit and Joseph Piatt

13.1 Introduction

Microcephalic primordial dwarfism is a rare group of single-gene disorders associated with extreme growth failure before and after birth, microcephaly, and skeletal dysplasia in addition to distinctive facial features. This disorder affects growth in utero and leads to severe small stature and microcephaly [1]. In this chapter, we describe multisystemic involvement of microcephalic primordial dwarfism and relevant features for the anesthesiologist providing anesthesia or sedation care to these individuals.

Patients with severe intrauterine growth retardation, microcephaly, and mental retardation with the facial features of receding forehead and chin and large beaked nose were described by Majewski et al. in 1982 in a series of three consecutive papers published in *American Journal of Medical Genetics*; these characteristics were defined as Majewski osteodysplastic primordial dwarfism types I, II, and III [2–4]. In 2004, Dr. Judith Hall et al. comprehensively reviewed the natural history of Majewski osteodysplastic primordial dwarfism type II in a large case series of 58 affected individuals [5]. Microcephaly is a distinguishing feature from other forms of dwarfism such as achondroplasia and mucopolysaccharidosis.

D. Dixit (✉)
Department of Anesthesiology and Perioperative Medicine, Nemours Children's Hospital, Wilmington, Delaware, USA

Anesthesiology and Pediatrics, Sidney Kimmel Medical College, Thomas Jefferson University, Philadelphia, Pennsylvania, USA
e-mail: Divya.Dixit@nemours.org

J. Piatt
Department of Neurosurgery, Nemours Children's Hospital, Wilmington, Delaware, USA
e-mail: Joseph.Piatt@nemours.org

M. C. Theroux (ed.), *Anesthesia Care of Patients with Skeletal Dysplasia*,
https://doi.org/10.1007/978-3-031-92501-6_13

The term microcephalic primordial dwarfism is used to describe a related group of disorders that are characterized by severe prenatal and postnatal growth failure together with microcephaly [1, 6]. Majewski et al. originally described primordial dwarfism as microcephalic osteodysplastic primordial dwarfism (MOPD) types I, II, and III. However, it is now understood that types I and III are the phenotypic variants of the same entity and are grouped together as MOPD I/III. Microcephalic osteodysplastic primordial dwarfism type II is the most common type and the most described form of microcephalic primordial dwarfism. Hence, for the purpose of this chapter, we shall describe MOPD II.

13.2 Genetics

Microcephalic osteodysplastic primordial dwarfism type II is transmitted in an autosomal recessive manner as a result of biallelic mutations causing loss of function in the pericentrin *(PCNT)* gene located on chromosome 21 [7]. This gene codes for a large protein localized to the centrosome. The centrosome organizes proper chromosome segregation during cell division, and it is thought that cells of individuals with MOPD II are more susceptible to defective mitosis and chromosome segregation, resulting in restriction of growth.

13.3 Phenotypic and Musculoskeletal Features

The distinguishing features of MOPD II are the characteristic facies, hypodontia, skeletal dysplasia, cerebrovascular disease, insulin resistance, and hypertension [6, 8]. Intrauterine growth retardation, exhibiting growth deficiency that progressively becomes more severe during the course of pregnancy, can be recognized with ultrasound between weeks 12 and 14 of pregnancy. In a study of 26 patients by Bober et al., the average gestational age at birth in patients with molecularly diagnosed MOPD II was 34.8 weeks ±2.8 weeks, with prematurity being common. The average length, weight, and head circumference at birth when compared with an otherwise healthy neonate are presented in Table 13.1 [9]. At skeletal maturity, the average height was 10.3 standard deviations (SD) below the mean, weight 14.3 SD below the mean, and head circumference 8.5 SD below the mean (Table 13.2). The final adult height of individuals with MOPD II is often reduced to about 100 cm [5]. Weight gain in individuals with MOPD II is poor and dramatically different compared with weight gain in typical infants and children [10]. Patients with MOPD II have been treated with growth hormone but appear to have little response to growth hormone therapy [9].

Table 13.1 Average length, weight, and head circumference of term infants with MOPD II at birth in comparison to a healthy neonate [9]

Child with MOPD II at birth	Compares to average neonate
Average birth weight	31–32-week neonate
Average birth length	28–29-week neonate
Average birth head circumference	30–31-week-old neonate

MOPD microcephalic osteodysplastic primordial dwarfism

Table 13.2 Average height, weight, and head circumference of skeletally mature males with MOPD II in comparison to a male with average stature [9]

Skeletally mature male with MOPD II	Compares to average male
Average height	3-year 10-month-old boy
Average weight	5-year 2-month-old boy
Average head circumference	5-month-old infant

MOPD microcephalic osteodysplastic primordial dwarfism

Patients with MOPD II have distinct craniofacial features [5, 6]. These include a prominent nose with a wide elevated nasal bridge. The mouth, though appearing small, is in proportion to the size of the face as is the tongue, which seems proportionate to the mouth. Micrognathia and a high arched palate frequently are present. Eyes appear prominent during childhood and there may be a mild down slanting to the palpebral fissures. Ears may be simple or dysplastic with attached lobules. The intelligence level of individuals with MOPD II is normal or near normal [5]. Hypodontia is another distinct feature of patients with MOPD II with increased spacing between primary teeth. Secondary teeth are small and dysplastic with hypoplasia of enamel and short or absent roots. This poor dentition leads to difficulty in chewing and loss of teeth by the second or third decade [5].

Generalized bony dysplasia affects epiphyses as well as metaphyses. The long bones are generally thin and gracile appearing with widening of the metaphyses. Dislocation or subluxation of the radial head can occur due to epiphyseal ossification delay. Mesomelia of the upper extremities develops with distal flaring of the radius and ulnar metaphyses. Wrist joints are knobby with ulnar deviation, and fifth digit clinodactyly may be present. The distal femoral epiphyses with ossification become triangular, and the distal femoral metaphyses develop an inverted V-shape (Fig. 13.1). While both coxa vara and coxa valga have been described, a study by Karatas et al. in 2014 examined the hip pathology in children with MOPD II and reported that 54% of hips were diagnosed with developmental coxa vara (three out of four unilateral), hip dysplasia, and hip subluxation [11]. Age at diagnosis of coxa vara is between 2.5 and 5 years, with some children requiring surgical in situ

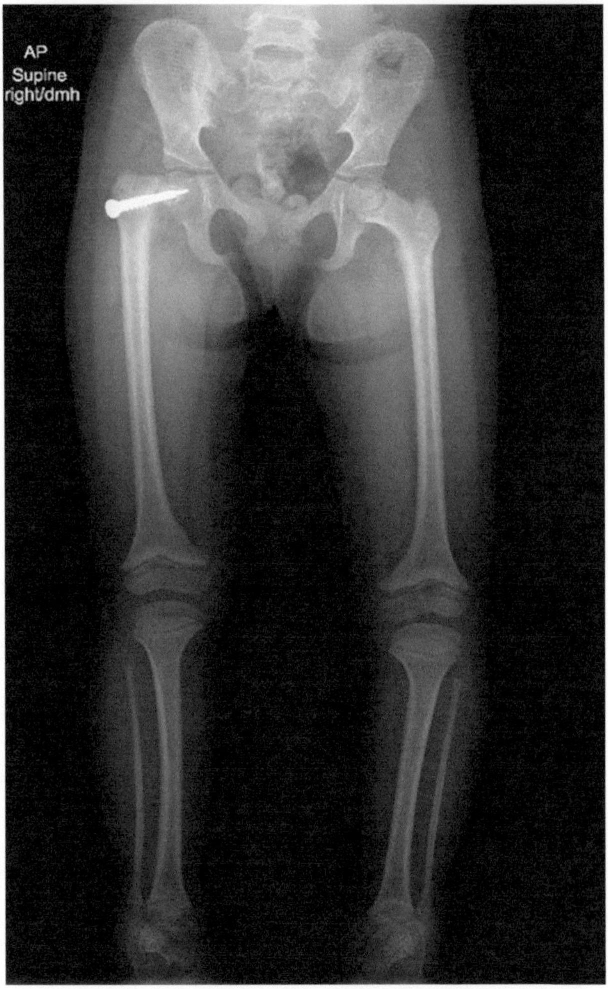

Fig. 13.1 Hip dysplasia: x-ray image of child with MOPD II. Anteroposterior supine view of pelvis and lower extremities in a 4-year-old child with MOPD II showing bilateral coxa vara, screw fixation of right slipped capital femoral epiphysis, inverted V-shaped distal femoral metaphysis, gracile long bones, shallow right acetabulum and narrow right hip joint space, and pelvic tilt with the right hip iliac crest higher. *MOPD* microcephalic osteodysplastic primordial dwarfism

pinning to stabilize the coxa vara. Slipped capital femoral epiphysis has also been described (Fig. 13.1) [12]. Routine clinical and radiologic evaluation of the hip is important to screen for and follow-up hip pathology. As children with MOPD II enter puberty, scoliosis, with or without kyphosis, can progress rapidly, sometimes requiring spinal fusion (Fig. 13.2).

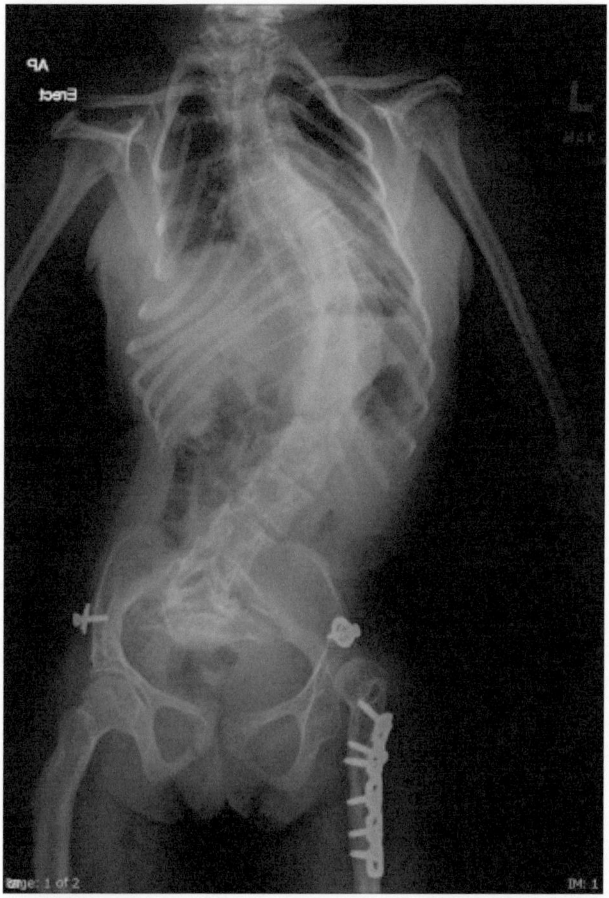

Fig. 13.2 Scoliosis: x-ray image of a child with MOPD II. X-ray scoliosis erect view in a 16-year-old female with MOPD II showing surgical fixation of left proximal femur and bilateral iliac bones, severe thoracolumbar levocurvature of 77°, additional dextrocurvatures in the cervical, upper thoracic and lumbosacral spine, and bell-shaped distorted chest with low lung volumes. *MOPD* microcephalic osteodysplastic primordial dwarfism

13.4 Central Nervous System

Cerebrovascular disease is prevalent in MOPD II and is a significant cause of morbidity. Stroke can lead to disability and even mortality. MOPD II is unique among the various types of skeletal dwarfism for anomalies of cerebral nervous system vasculature—most distinctively moyamoya syndrome [13, 14]. "Moyamoya" is a Japanese expression often translated as "puff of smoke." It refers to the angiographic appearance of collateral blood vessels from the vertebrobasilar system that develops in response to progressive stenosis of the supraclinoid internal carotid arteries and their major branches by fibrocellular intimal thickening [15]. Collaterals may develop from the external carotid system as well. In an unusual but illustrative case in a child with MOPD II, presumed involvement of the origin of the ophthalmic artery led to spontaneous revascularization of the globe with a prominent collateral vessel passing across the iris [16]. The usual clinical presentation of moyamoya is ischemic cerebral infarction, but hemorrhage can occur as well from microaneurysms arising from the fragile collateral vessels. True saccular aneurysms may occur as well [15]. Eleven of the 58 individuals reported by Hall et al. in a review of natural history study of MOPD II had cerebral vascular involvement, and of these 11 patients, 4 (19%) died from rupture of vessels between the ages of 13 and 25 years. In the same study, strokes were reported in younger children at age 3 years or less [5]. Based on these data, the authors suggested that patients with MOPD II should be screened routinely for the development of intracranial vascular abnormalities [5]. In a clinical report and review, Brancati et al. documented that 15 of 63 (23.8%) patients were found to have cerebral aneurysms or moyamoya disease [14]. In a study of 47 individuals with confirmed MOPD II, Duker et al. reported 47% were diagnosed with moyamoya, 53% had intracranial aneurysms, and 36% were diagnosed with both moyamoya and intracranial aneurysms [8]. More than one aneurysm were detected in 16 individuals in this study. Perry et al. reported 32% of 147 patients with MOPD II had cerebrovascular pathology, occlusive arteriopathy (stenosis, occlusion), and a significant association with the younger age group when compared with aneurysmal cerebrovascular disease (Fig. 13.3) [17].

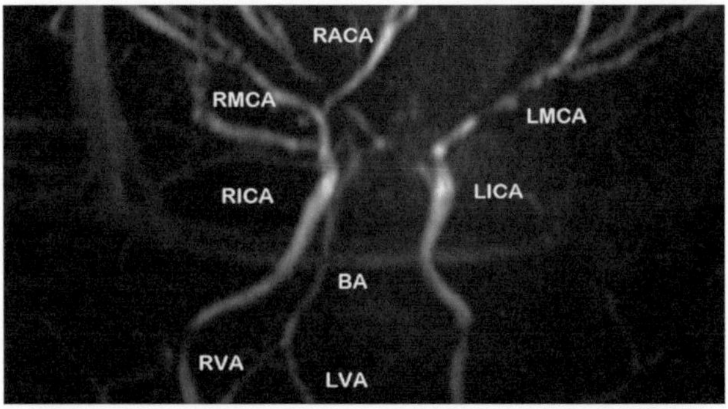

Fig. 13.3 Cerebrovascular arteriopathy/moyamoya in a child with MOPD II. Magnetic resonance angiography in a 3-year 8-month-old child with MOPD II showing cerebrovascular occlusive pathology. Both internal carotid arteries are narrowed just proximal to their bifurcations. The right middle and anterior cerebral arteries are extremely narrowed. The left middle cerebral artery is narrowed with a beaded appearance. The left anterior cerebral artery is not visible at all. Both vertebral arteries and the basilar artery are narrowed. None of the major cerebral arteries are normal in caliber. *BA* basilar artery, *LICA* left internal carotid artery, *LMCA* left middle cerebral artery, *LVA* left vertebral artery, *MOPD* microcephalic osteodysplastic primordial dwarfism, *RACA* right anterior cerebral artery, *RICA* right internal carotid artery, *RMCA* right middle cerebral artery, *RVA* right vertebral artery

Table 13.3 Neurosurgical procedures performed in patients with MOPD II

Direct vascularization
Superficial temporal artery to middle cerebral artery bypass
Indirect revascularization
Encephalo-duro-arterio-synangiosis
Encephalo-myo-synangiosis
Omental-cerebral transposition
Aneurysmal surgery
Endovascular coil occlusion
Endovascular stenting
Microsurgical clipping

MOPD microcephalic osteodysplastic primordial dwarfism

The definitive diagnostic investigation for moyamoya angiopathy has been cerebral arteriography, which remains indispensable in the management of true cerebral aneurysms. Magnetic resonance imaging (MRI) and magnetic resonance angiography (MRA), however, are the screening tests of choice and play a growing role in clinical management. Acute and chronic cerebral infarctions are readily visualized. Slow cortical arterial flow can be detected as the "ivy sign" on fluid-attenuated inversion recovery MRI sequences [18]. Static disturbances of cerebral perfusion are evident on arterial spin labeling, and stress-testing with acetazolamide allows assessment of cerebral arterial reserve. The MRA captures stenoses of the major cerebral arteries and the collateral vessels for which the condition moyamoya is named. Screening ought to be initiated early in childhood. Whether angiopathy is present at birth is unknown, but it is predictably progressive [15]. Moyamoya appears earlier in childhood in MOPD II, while aneurysms have been diagnosed throughout the lifespan [19]. Any change in neurological status such as speech difficulties, changes in motor status and cognition, headache, and seizures may warrant neurological examinations and studies to rule out disease progression [17]. Screening of asymptomatic individuals with MOPD II has been recommended every 12–18 months [8, 15, 17].

The cerebral angiopathy associated with MOPD II is treatable [19–21]. Neurosurgical interventions are presented in Table 13.3. Revascularization procedures are undertaken to prevent cerebral infarction, but by unloading fragile collateral vessels, these procedures may suppress the risk of intracerebral hemorrhage as well. There are two paradigms: (1) direct revascularization entails surgical anastomosis of a scalp artery, typically the superficial temporal artery, to a major branch of the middle cerebral artery [21]; and (2) arteries in younger children are small, so an alternative indirect approach sometimes is needed. Indirect revascularization relies on powerful angiogenic factors produced by the ischemic brain tissue to stimulate the in-growth of collateral arteries from vascularized tissues laid on the cortical surface [19]. Candidate donor tissues include the transposed superficial temporal artery, inverted dura, temporalis muscle, and even free grafts of omentum. The largest surgical experience with treatment of the cerebral angiopathy of patients with MOPD II is the report of ten patients from the Steinberg group at Stanford University Medical Center [19]. As noted previously, moyamoya presented earlier in childhood and predominantly was treated with indirect methods. Aneurysms were coiled or clipped conventionally. One patient died of subarachnoid hemorrhage from a new untreated aneurysm, but the neurological condition of the survivors was stabilized.

13.5 Other Co-morbid Conditions

Subglottic stenosis and laryngomalacia are prevalent in children with MOPD II. A relatively high-pitched squeaky nasal voice is quite characteristic in these children and is probably related to a narrow larynx and trachea. Subglottic stenosis has been described in four individuals, with two individuals developing severe disease requiring tracheostomy [5]. No significant cardiac abnormalities are expected in MOPD II; however, in a landmark paper, Hall et al. reported bicuspid aortic valve ($n = 1$), atrial septal defect (ASD) ($n = 1$), and patent ductus arteriosus (PDA) ($n = 2$) in their study of 58 affected individuals [5]. In addition, a case of myocarditis and myocardial infarct in an affected female with MOPD II was also reported in this study. Blood pressure (BP) was generally considered normal [5]. However, the study by Duker et al. in 2021 noted that many of the individuals (43% in a cohort of 47 individuals) were diagnosed with hypertension and were on antihypertensive medications [8]. In that study, the cut-off measurement for hypertension was a BP of 110/70 mmHg corresponding to the 95th percentile for males and females at this height, as described by Flynn et al. [22]. Sixty-six percent of MOPD II patients (31 individuals) had one documented BP higher than 120/80 mmHg [8]. Seventy-seven percent of individuals with moyamoya disease, 84% with aneurysms, and 75% with myocardial infarction had BP measurements greater than 110/70 mmHg [8]. In the absence of specific criteria, it is recommended that BP higher than 120/80 not be considered normal for individuals with MOPD II [8]. The same study also described 28% of individuals with MOPD II diagnosed with ASD, ventricular septal defect (VSD), or patent foramen ovale (PFO) beyond the newborn period, and one individual had multiple rhabdomyomas. In the presence of hypertension, electrocardiogram and echocardiogram may be considered.

Other system abnormalities described in children with MOPD II include renal artery stenosis and renal artery aneurysms. Chronic kidney disease has been diagnosed in 32% of cohorts ($n = 47$), and 88% of individuals with chronic kidney disease of stage 3 or greater had BP higher than 110/70 mmHg [8]. Renal transplantation and nephrolithiasis have been described in two and six individuals, respectively. Renal abnormalities such as accessory/duplicated renal arteries, hydronephrosis, renal stones, and malposition and underdevelopment have also been described. Renal ultrasound has been recommended at diagnosis as well as an assessment of renal function starting at age 5 years [23]. Cryptorchidism or retractile testes have been described in up to 44% of males with MOPD II, and hypospadias has been diagnosed in some individuals [8]. Some of these children may present for surgical interventions such as orchiopexy.

Diabetes and/or insulin resistance were present in almost one-third of the individuals with MOPD II (18 of 47) in one study [8]. Insulin resistance is described at

Table 13.4 Common causes of mortality in individuals with MOPD II [8]

Cause of death	Number of individuals	Age of death (in years)
Ruptured brain aneurysm	5	7–24
Respiratory failure	1	30
Myocardial infarction/coronary artery disease	3	18–25
Heart/multiorgan failure after surgery	1	30
Unknown causes (2 had history of moyamoya revascularization surgery, 1 had kidney failure and prior myocardial infarction)	3	13–41

MOPD microcephalic osteodysplastic primordial dwarfism

an early age in a different study, commonly manifesting between ages 5 and 10 years in children with MOPD II [24]. These individuals should be followed regularly for metabolic function, including fasting glucose, lipid profile, and liver function [23].

Dermatological manifestations such as acanthosis nigricans, flexural hyperpigmentation, and skin tags have been documented and are likely related to insulin resistance [5, 24]. Café-au-lait spots are a common dermatological lesion described in children with MOPD II [13, 25]. Wizening of hands with multiple skin creases have been noted with aging. Cutis marmorata and mottling have also been described [14].

Anemia was a frequent in the study by Duker et al., described in 93% of individuals (hemoglobin <12 g/dL), as were hypercholesterolemia (>200 mg/dL; 58%) and thrombocytosis (90%) [8]. Atrophic macular scarring is described in three patients who had normal retinal pigmentation [26].

With early intervention and greater understanding of the disease, individuals with MOPD are living longer. However, the lifespan of these individuals remains shortened, with the oldest living individual in their 40s now. Thirteen deaths have been recorded in individuals with MOPD II, and causes of death are summarized in Table 13.4.

13.6 Anesthesia Care

As in many forms of skeletal dysplasia, children with MOPD II require medical imaging studies such as MRI and MRA (for screening and monitoring for neurovascular anomalies), which bring them to the attention of the anesthesiology department. The various procedures performed in children with MOPD in our institution are listed in Table 13.5.

Thorough history and complete physical examination with special attention to airway evaluation and any past difficulty with tracheal intubation remain essential to providing safe anesthetic care. Special attention should be paid to BP measurements as up to 49% of individuals with MOPD could be hypertensive or prehypertensive [8]. Basic laboratory evaluation should include complete blood count to rule out anemia and basic metabolic panel to rule out prediabetic state. Considering the small thoracic cavities and limited pulmonary capacity of children with MOPD II,

Table 13.5 Procedures performed in children with MOPD I/III and MOPD II at our institution

Type of surgery	Procedure	Indication
General surgery	Gastrostomy tube placement	Aspiration risk, improvement in nutrition
	Nissen fundoplication	Gastroesophageal reflux
Gastroenterology procedure	Upper gastrointestinal endoscopy	Gastroesophageal reflux
Orthopedic surgery	Posterior spinal fusion	Scoliosis
	Hip pinning, arthrogram	Slipped capital femoral epiphyses
	Proximal femoral osteotomy and blade plate fixation	Hip dysplasia
Genitourinary surgery	Orchiopexy	Cryptorchidism
	Inguinal herniorrhaphy	Inguinal hernia
	Cystourethroscopy, stent placement, pyeloplasty	Ureteropelvic junction obstruction
Otorhinolaryngology surgery	Myringotomy tube	Otitis media
	Tracheostomy	Difficult airway
Dental surgery	Dental rehabilitation	Poor dentition
Neurosurgery	Ventriculoperitoneal shunt	Hydrocephalus
Neuroimaging studies	MRI, MRA	Cerebrovascular anomalies

MOPD microcephalic osteodysplastic primordial dwarfism, *MRI* magnetic resonance imaging, *MRA* magnetic resonance angiography,

smaller tidal volume may be indicated while on positive pressure ventilation. However, as these individuals are small in their body habitus, weight-based tidal volume is often adequate. While major cardiac abnormalities are rare, the more common defects such as ASD, VSD, PFO, and PDA should be suspected, and, if indicated, echocardiogram should be performed.

Difficulty with endotracheal intubation is often encountered. Such difficulties may include visualizing and accessing the larynx or advancing the endotracheal tube once the endotracheal tube is introduced through the larynx. It is therefore essential to prepare for difficult intubation prior to inducing general anesthesia in MOPD II patients. A smaller than anticipated (based on age) endotracheal tube is the norm and should not be surprising given the small stature and weight of the patient. The downsized endotracheal tube is usually sufficient to ventilate the patient due to the smaller size and weight; in other words, the amount of tissue that needs to be adequately ventilated and oxygenated is proportionately small. One of our patients with MOPD had a documented history of cardiorespiratory arrest that occurred during the induction of anesthesia, and another required an emergent tracheostomy. It may be prudent to be prepared with alternate means of tracheal intubation other than direct laryngoscopy. This could include flexible bronchoscope, video laryngoscope, and supraglottic devices such as the laryngeal mask airway.

When moyamoya vasculopathy is present in a patient with MOPD II, anesthetic management poses additional risks. Rupture of dilated, fragile moyamoya vessels or rupture of saccular aneurysm in the circle of Willis may be induced by hemodynamic stress. Cerebral ischemia may be induced by hyperventilation leading to reduced hemispheric blood flow [27]. Therefore, the anesthetic technique must focus on maintaining adequate cerebral blood flow, normalization of intracranial pressure, and avoidance of both cerebral vasoconstriction and vasodilatation. Temperature should be maintained within normal ranges, avoiding both hypo- and hyperthermia. It is critical that adequate hydration and normocapnia be maintained throughout the procedure [28]. Iwama et al. analyzed 236 cerebrovascular reconstruction surgeries in 124 patients younger than 15 years with moyamoya disease [28]. In this study, 21 children experienced perioperative ischemic events despite prevention of hypocapnia and hypotension. However, the Iwama et al. study did find that the incidence of hypercapnia (maximum $PaCO_2 > 45$ mmHg) was significantly higher intra- and postoperatively in patients with perioperative ischemic complications, emphasizing maintenance of normocapnia [28]. Avoiding large swings in BP by minimizing coughing and bucking around the endotracheal tube is ideal. However, achieving these goals in a patient with a potentially difficult airway may be daunting.

Because of the small patient size, body weight, and significant blood loss associated with major surgical procedures such as spinal fusion, accurate assessment of blood loss is essential, and replacement should be prompt so that excessive crystalloid administration may be avoided. Emphasis should be placed on minimizing airway swelling and safe extubation of the trachea at the end of the surgical procedure.

In general, neuraxial anesthesia is avoided in patients with skeletal dysplasia given that there are a large number of classes of skeletal dysplasia, each with different degrees of spinal abnormalities. Abnormalities of the spine, such as spinal canal stenosis, may not be evident in a young patient with MOPD. There is a theoretical concern of spinal stenosis as described in various other skeletal dysplasia syndromes; peripheral nerve blocks may be employed when appropriate, which may aid in minimizing sedatives and analgesics, especially narcotics.

Nature of the surgery, preoperative comorbidity, and difficulty encountered during intraoperative course influence the postoperative management of these patients. It is best to attempt to extubate the trachea in the operating room at the end of the procedure and determine the need for supplemental ventilatory assistance such as use of high-flow nasal canula (Vapotherm, Exeter, NH, USA). Admission to the intensive care unit (ICU) may be necessary and should be determined on a case-by-case basis. Need for reintubation should be a concern and such a scenario might be discussed at the time of patient drop off in the ICU. Preplanning could avoid unnecessary delay and possible harm to the patient. Considering many of the children with MOPD require a smaller than anticipated endotracheal tube size, postoperative stridor is a possibility, which could be treated using racemic epinephrine. Closer monitoring of BP is necessary for revascularization procedures, and in most situations, postoperative ICU admission is required.

13.7 Case Illustration

A 13-year-old, 8.2-kg female diagnosed with microcephalic primordial dwarfism and progressive thoracic kyphoscoliosis (Cobb's curve of 90°) was scheduled for anterior thoracic release and posterior spinal fusion with instrumentation. History and examination revealed a cognitively normal child with strikingly small stature (95.25 cm), abnormal facial profile with micrognathia, and large beaked nose. She had significant malocclusion of the jaw, a loose upper right molar, and Mallampati class 3 airway. Her single previous anesthetic for a dental procedure had resulted in failed endotracheal intubation, but mask ventilation was feasible. She had restrictive lung disease with a small asymmetrical chest, with bilateral good air entry and unlabored respirations. Cardiac examination revealed normal cardiac function. Routine blood work was within normal limits. Neuroimaging of the brain did not indicate any structural or vascular abnormality. An MRI of the spine indicated severe thoracic kyphosis with no spinal cord abnormality. Electrocardiogram and echocardiogram indicated normal sinus rhythm and cardiac anatomy.

Preparations were made for a potentially difficult endotracheal intubation. Following premedication with intranasal midazolam, she was taken to the operating room. Routine monitors and an intravenous line were placed. Anesthesia was induced with glycopyrrolate and 8% sevoflurane in oxygen; once the possibility of easy mask ventilation was confirmed, rocuronium was administered. In view of clearly difficult laryngoscopy and loose teeth, no attempt was made to perform direct laryngoscopy. Following unsuccessful attempts using a Bonfils intubating stylet (Storz, Tuttlingen, Germany), flexible bronchoscopy-aided intubation was successful using the nasal route. (Of note, we are no longer using the intubating stylet with the advent of video laryngoscopes and greater experience using flexible bronchoscope). Thus, a 4.5-mm internal diameter cuffed endotracheal tube was placed. No further difficulty with airway or ventilation was experienced until toward the end of surgery as noted below. Two intravenous lines, a central venous access and radial arterial access, and a urinary catheter were placed prior to the beginning of surgery. The inhalational agent was then discontinued, and anesthesia was maintained primarily with propofol and remifentanil infusion. Evoked potential monitoring was established and the patient was positioned initially in the left lateral decubitus position for anterior thoracic spine release and subsequently in the prone position for the posterior spinal fusion. The procedure lasted for approximately 12 h with 1-L blood loss. Replacement for blood loss included packed red blood cells, fresh frozen plasma, and platelets.

In the later part of the surgical procedure, the patient required higher than normal ventilatory pressures. Frothy, blood-tinged secretions were noted in the endotracheal tube, indicative of pulmonary edema that responded to 5 mg of intravenous furosemide. Dopamine infusion was commenced at the time at 5 mcg/kg/min. The remainder of the intraoperative course was uneventful.

Following completion of the surgical procedure, the patient was transferred to the pediatric ICU intubated and on mechanical ventilation. On the fourth postoperative day, the decision was made to extubate the trachea. However, within a few hours following tracheal extubation, the patient developed progressive worsening of respiratory status. Bi-level positive airway pressure with a facemask provided no relief and the decision was made to reintubate the trachea. Following multiple attempts at direct laryngoscopy by ICU physicians and anesthesiologists, reintubation was accomplished using a flexible bronchoscope via the nostril. The patient remained intubated for 4 more days and was successfully extubated and finally discharged from the hospital after 16 days.

This case illustrates pitfalls we have encountered, especially regarding difficult intubation, and the scenario of requiring reintubation. Among patients with microcephalic primordial dwarfism, those with MOPD II most frequently present for a scheduled procedure under anesthesia and pose significant challenges to the anesthesiologist. These challenges relate to tenuous airway, poor respiratory reserve, cerebrovascular pathology, skeletal dysplasia, renal abnormalities, and poor dentition. Managing a difficult airway in the face of moyamoya-related vasculopathy can be an unusual challenge. Thorough clinical and laboratory assessment, radiologic imaging, careful planning, and skilled conduct of anesthesia are essential for uneventful anesthetic administration and a favorable outcome. These patients pose a substantial risk for anesthesiologists, and we are hopeful that this chapter detailing our experiences will be helpful in providing care for these patients.

References

1. Klingseisen A, Jackson AP. Mechanisms and pathways of growth failure in primordial dwarfism. Genes Dev. 2011;25:2011–24.
2. Majewski F, Goecke T. Studies of microcephalic primordial dwarfism I: approach to a delineation of the Seckel syndrome. Am J Med Genet. 1982;12:7–21.
3. Majewski F, Ranke M, Schinzel A. Studies of microcephalic primordial dwarfism II: the osteodysplastic type II of primordial dwarfism. Am J Med Genet. 1982;12:23–35.
4. Majewski F, Stoeckenius M, Kemperdick H. Studies of microcephalic primordial dwarfism III: an intrauterine dwarf with platyspondyly and anomalies of pelvis and clavicles—osteodysplastic primordial dwarfism type III. Am J Med Genet. 1982;12:37–42.
5. Hall JG, Flora C, Scott CI Jr, Pauli RM, Tanaka KI. Majewski osteodysplastic primordial dwarfism type II (MOPD II): natural history and clinical findings. Am J Med Genet A. 2004;130A:55–72.
6. Bober MB, Jackson AP. Microcephalic osteodysplastic primordial dwarfism, type II: a clinical review. Curr Osteoporos Rep. 2017;15:61–9.
7. Rauch A, Thiel CT, Schindler D, Wick U, Crow YJ, Ekici AB, et al. Mutations in the pericentrin (PCNT) gene cause primordial dwarfism. Science. 2008;319:816–9.
8. Duker AL, Kinderman D, Jordan C, Niiler T, Baker-Smith CM, Thompson L, et al. Microcephalic osteodysplastic primordial dwarfism type II is associated with global vascular disease. Orphanet J Rare Dis. 2021;16:231.
9. Bober MB, Niiler T, Duker AL, Murray JE, Ketterer T, Harley ME, et al. Growth in individuals with majewski osteodysplastic primordial dwarfism type II caused by pericentrin mutations. Am J Med Genet A. 2012;158A:2719–25.
10. Duker AL, Niiler T, Bober MB. Expected weight gain for children with microcephalic osteodysplastic primordial dwarfism type II. Am J Med Genet A. 2017;173:3067–9.
11. Karatas AF, Bober MB, Rogers K, Duker AL, Ditro CP, Mackenzie WG. Hip pathology in majewski osteodysplastic primordial dwarfism type II. J Pediatr Orthop. 2014;34:585–90.

12. Majewski F, Goecke TO. Microcephalic osteodysplastic primordial dwarfism type II: report of three cases and review. Am J Med Genet. 1998;80:25–31.
13. Nishimura G, Hasegawa T, Fujino M, Hori N, Tomita Y. Microcephalic osteodysplastic primordial short stature type II with cafe-au-lait spots and moyamoya disease. Am J Med Genet A. 2003;117A:299–301.
14. Brancati F, Castori M, Mingarelli R, Dallapiccola B. Majewski osteodysplastic primordial dwarfism type II (MOPD II) complicated by stroke: clinical report and review of cerebral vascular anomalies. Am J Med Genet A. 2005;139:212–5.
15. Bober MB, Khan N, Kaplan J, Lewis K, Feinstein JA, Scott CI Jr, et al. Majewski osteodysplastic primordial dwarfism type II (MOPD II): expanding the vascular phenotype. Am J Med Genet A. 2010;152A:960–5.
16. Bang GM, Kirmani S, Patton A, Pulido JS, Brodsky MC. "Ocular moyamoya" syndrome in a patient with features of microcephalic osteodysplastic primordial dwarfism type II. J AAPOS. 2013;17:100–2.
17. Perry LD, Robertson F, Ganesan V. Screening for cerebrovascular disease in microcephalic osteodysplastic primordial dwarfism type II (MOPD II): an evidence-based proposal. Pediatr Neurol. 2013;48:294–8.
18. Mori N, Mugikura S, Higano S, Kaneta T, Fujimura M, Umetsu A, et al. The leptomeningeal "ivy sign" on fluid-attenuated inversion recovery MR imaging in moyamoya disease: a sign of decreased cerebral vascular reserve? AJNR Am J Neuroradiol. 2009;30:930–5.
19. Teo M, Johnson JN, Bell-Stephens TE, Marks MP, Do HM, Dodd RL, et al. Surgical outcomes of majewski osteodysplastic primordial dwarfism type II with intracranial vascular anomalies. J Neurosurg Pediatr. 2016;25:717–23.
20. Waldron JS, Hetts SW, Armstrong-Wells J, Dowd CF, Fullerton HJ, Gupta N, et al. Multiple intracranial aneurysms and moyamoya disease associated with microcephalic osteodysplastic primordial dwarfism type II: surgical considerations. J Neurosurg Pediatr. 2009;4:439–44.
21. Guzman R, Lee M, Achrol A, Bell-Stephens T, Kelly M, Do HM, et al. Clinical outcome after 450 revascularization procedures for moyamoya disease. Clinical article. J Neurosurg. 2009;111:927–35.
22. Flynn JT, Kaelber DC, Baker-Smith CM, Blowey D, Carroll AE, Daniels SR, et al. Clinical practice guideline for screening and management of high blood pressure in children and adolescents. Pediatrics. 2017;140:e20171904.
23. Duker A, Jackson A, Bober MB. Microcephalic osteodysplastic primordial dwarfism type II. In: Adam MP, Mirzaa GM, Pagon RA, Wallace SE, LJH B, Gripp KW, et al., editors. GeneReviews® [Internet]. Seattle: University of Washington, Seattle; 1993–2024.
24. Huang-Doran I, Bicknell LS, Finucane FM, Rocha N, Porter KM, Tung YCL, et al. Genetic defects in human pericentrin are associated with severe insulin resistance and diabetes. Diabetes. 2011;60:925–35.
25. Kannu P, Kelly P, Aftimos S. Microcephalic osteodysplastic primordial dwarfism type II: a child with café au lait lesions, cutis marmorata, and moyamoya disease. Am J Med Genet A. 2004;128A:98–100.
26. Chen WJ, Huang FC, Shih MH. Ocular characteristics in a variant microcephalic primordial dwarfism type II. BMC Pediatr. 2019;19:329.
27. Takeuchi S, Tanaka R, Ishii R, Tsuchida T, Kobayashi K, Arai H. Cerebral hemodynamics in patients with moyamoya disease. A study of regional cerebral blood flow by the 133Xe inhalation method. Surg Neurol. 1985;23:468–74.
28. Iwama T, Hashimoto N, Yonekawa Y. The relevance of hemodynamic factors to perioperative ischemic complications in childhood moyamoya disease. Neurosurgery. 1996;38:1120–5; discussion 1125–6.

Morquio A (MPS IVA) Syndrome

14

Mary C. Theroux, Abraham Oommen, Sabina Dicindio, and Lauren W. Averill

14.1 Introduction

Mucopolysaccharidosis type IVA (MPS IVA) is a lysosomal storage disease characterized by the accumulation of glycosaminoglycans (GAGs) in multiple tissues—particularly bone, cartilage, heart valves, and cornea [1]. Glycosaminoglycans mainly consist of keratan sulfate (KS) and chondroitin-6-sulfate (C6S), and their accumulation leads to progressive organ dysfunction, impacting respiratory, cardiac, and skeletal systems including the spinal cord [2, 3]. Vision and hearing are affected as well [4]. The co-morbid conditions impact anesthesia care, most importantly impacting airway management and spinal cord protection. Airway management is difficult due to GAG deposits in supraglottic structures, resulting in a large tongue and narrowing of nostrils and other structures in close proximity to the larynx [5, 6]. Other factors contributing to difficult airway management are the short neck, pectus carinatum, and limited mouth opening [7, 8]. In addition to the obstruction of upper airway, a second level of obstruction occurs involving the trachea itself.

M. C. Theroux (✉) · A. Oommen · S. Dicindio
Department of Anesthesiology and Perioperative Medicine, Nemours Children's Hospital, Wilminton, Delaware, USA

Anesthesiology and Pediatrics, Sidney Kimmel Medical College, Thomas Jefferson University, Philadelphia, Pennsylvania, USA
e-mail: mtheroux@nemours.org; Abraham.oommen@nemours.org; sabina.dicindio@nemours.org

L. W. Averill
Radiology and Pediatrics, Sidney Kimmel Medical College, Thomas Jefferson University, Philadelphia, Pennsylvania, USA

Department of Radiology, Nemours Children's Hospital, Wilmington, Delaware, USA
e-mail: Lauren.averill@nemours.org

M. C. Theroux (ed.), *Anesthesia Care of Patients with Skeletal Dysplasia*,
https://doi.org/10.1007/978-3-031-92501-6_14

This is due to the narrowing, buckling, and twisting of the trachea that develops at the thoracic inlet in most patients with severe phenotype of Morquio A [9].

In patients with Morquio A, the spinal cord is vulnerable to ischemia and/or infarction during anesthesia, especially the T1–T7 segments [10, 11]. Deposits of GAGs lead to aortic and mitral valve disease and, less commonly, coronary artery disease. These co-morbid conditions warrant thorough evaluation to assess the severity and to provide optimal anesthetic management.

14.2 Genetics

Morquio syndrome is inherited in an autosomal-recessive pattern. It is important to recognize that Morquio syndrome has two types: Morquio A, known as MPS IVA (OMIM 253000), and Morquio B, known as MPS IVB (OMIM 253010). Morquio A results from a deficiency of the lysosomal enzyme N-acetylgalactosamine-6-sulfate sulfatase (GALNS), and Morquio B results from a deficiency of a different lysosomal enzyme, β-galactosidase. The human *GALNS* gene (which codes the enzyme deficient in Morquio A) is localized at 16q24.3 and is approximately 50 kb long, containing 13 introns and 14 exons [12, 13]. Four-hundred or more *GALNS* alleles have been identified, out of which an estimated 328 mutations have been shown to cause the disease phenotype. The ten most prevalent *GALNS* mutations account for 35% of all known cases of Morquio A. The correlation between genotype and phenotype may be useful in predicting clinical severity [14]. Morquio B syndrome is a different disease and should not be confused with Morquio A. Similarity between the names has led to confusion among patients and health care providers alike. A commonly observed misconception among health care providers and occasionally among patients is that Morquio B is an attenuated phenotype of Morquio A, which it is not.

14.3 Phenotype

Patients with MPS IVA typically appear healthy at birth but begin to show symptoms by age 3. Patients come to medical attention during the second or third year of life because of short stature (short-trunked), genu valgum, hypermobile joints, pectus carinatum, kyphoscoliosis, and odontoid hypoplasia (Fig. 14.1). Patients with the severe phenotype often do not survive beyond the second or third decade of life, primarily due to weakness of respiratory muscles and respiratory insufficiency originating from cervical stenosis and related spinal cord compromise [15]. In addition, progressive respiratory compromise from tracheal narrowing and buckling adds to the overall decline of the respiratory system [16]. Other causes of death are cardiac involvement and cardiac failure, which typically occurs in later age (late 20s and 30s) [17]. Coronary artery disease has been described by Factor et al., who reported intimal sclerosis as a prominent feature of Morquio A. They proposed the term "thesaurosclerosis" in order to differentiate its etiology as a lysosomal storage disease and avoid being confused with "atherosclerosis," related to lysosomal cholesterol storage

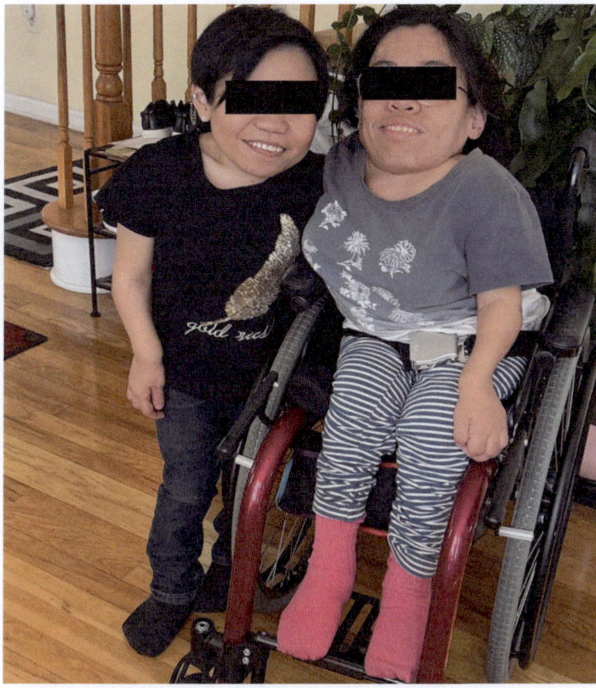

Fig. 14.1 Adult patients with Morquio A showing characteristic short-trunked stature

disease [18]. Vision impairment is common and starts with micro deposits of GAGs in the cornea, resulting in clouding [19] which ultimately requires surgical intervention including corneal transplant. Other ocular manifestations include retinal degeneration, decreased peripheral vision, night blindness, and glaucoma [20]. Progressive hearing impairment occurs and may require a cochlear implant [21, 22].

Severe pectus carinatum is common (Fig. 14.2) and may interfere with optimal positioning during surgical procedures in the prone position. During the placement of a central venous port to facilitate enzyme treatment therapy [23], the severe pectus deformity may interfere with optimally placing the port in the anterior chest wall. Dental abnormalities are common in Morquio A syndrome and include dental caries, tooth fractures, and abnormally thin enamel. Widely spaced teeth with thin enamel and frequent caries are often observed [24].

Tracheal abnormalities characterized by narrowing, buckling, and twisting may begin as early as 7 years of age in patients with Morquio A (Fig. 14.3). Patients often acquire a characteristic head and neck posture described as the "look up to the sky" position, where the head and neck are kept extended (Fig. 14.4). This posture is thought to help straighten the tracheal twisting and thus improve airway patency.

Clinical assessments to evaluate functional impact of the disease for patients with Morquio A are performed by 6-min walk, 3-min stair climb, and forced pulmonary tests. However, these endurance tests are difficult for small children,

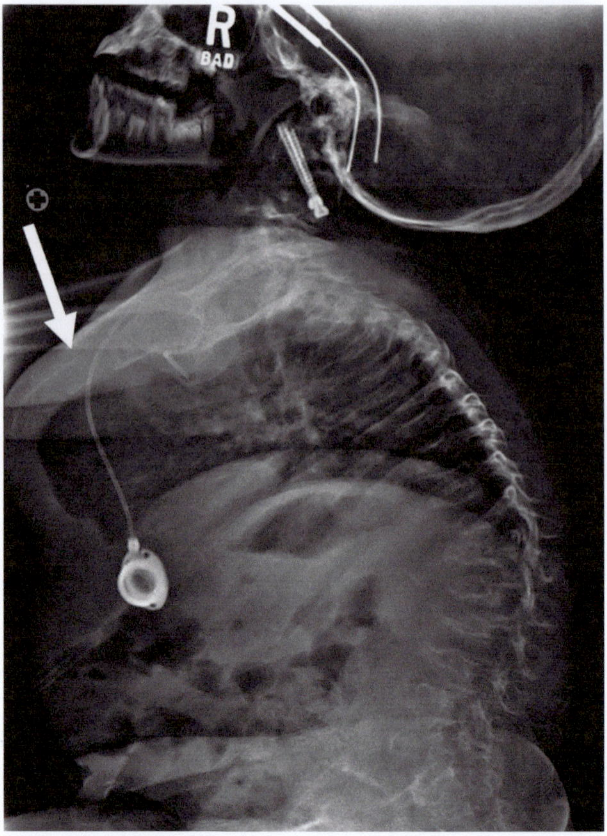

Fig. 14.2 Plain X-ray of the chest in lateral position of a 9-year-old girl with Morquio A syndrome. Pectus carinatum (arrow) is quite prominent in patients with Morquio A and poses challenges, especially when positioning the patients prone. Direct laryngoscopy may also be more difficult due to the prominent pectus carinatum. Note that the patient has a port for enzyme replacement therapy, and she has already undergone C1–C2 fusion

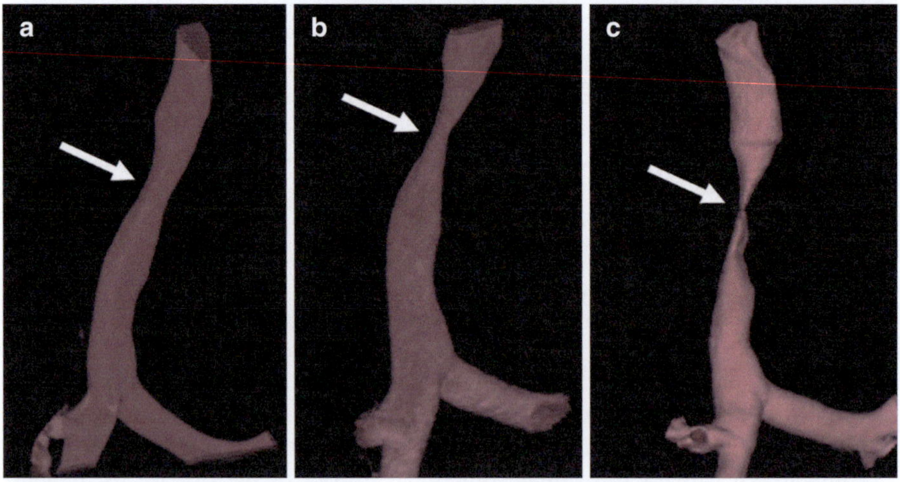

Fig. 14.3 Progressive tracheal narrowing in a boy with Morquio A syndrome. Three-dimensional reconstructions of the trachea from CT scans performed at 6- (**a**), 8- (**b**), and 14-years-old (**c**) show progressive narrowing of the trachea at the thoracic inlet (arrows). At the last time point, the tracheal cross-sectional area at the thoracic inlet was narrowed by 84% compared to the intrathoracic trachea

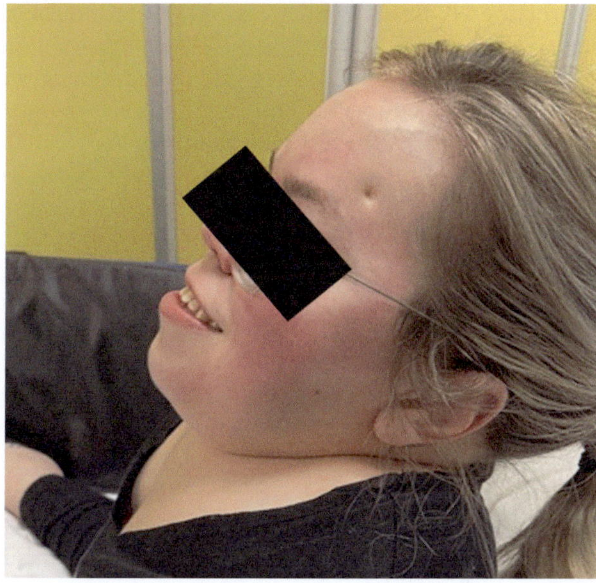

Fig. 14.4 Twenty-five-year-old patient with Morquio with characteristic posture of head and neck held in moderate extension. The indentation on the temple is from the pins used to hold the halo traction device following her cervical spine fusion at age 2.5 years. The halo was in place for 6 months following surgery

wheelchair-bound patients, and patients who have recently undergone orthopedic surgical procedures. There are ongoing studies designed to evaluate patients in a non-invasive way that may accommodate a broad spectrum of patients irrespective of age and physical disabilities.

14.4 Anesthesia Care

From an early age, patients with Morquio A require sedation and anesthesia for numerous orthopedic and other surgical procedures, including port placement for enzyme therapy. Several co-morbid conditions complicate the anesthesia and perioperative care of these patients, the most notable of which is related to the airway [5, 7]. Deposits of GAGs present through the airway passages (nostrils, oropharynx, tongue, larynx, and trachea) progressively cause narrowing resulting in moderate-to-severe upper airway obstruction. Obstruction at the level of trachea itself is unique to patients with Morquio A [9, 16, 25] and adds to the already compromised respiratory system. The severity of tracheal obstruction is age-related and occurs mostly between ages 10 and 20 years [9, 26]. The effect of the different modalities of treatment available (i.e., enzyme replacement therapy [ERT] or hematopoietic stem-cell transplantation) is unknown, but there is hope for a favorable outcome.

To enumerate, it is important to recognize that patients with Morquio A have airway difficulties related to different aspects of airway management. In 2015, we investigated a 16-year-old with Morquio A and severe upper airway obstruction whose CT angiogram (CTA) of the chest revealed an 80% obstruction of the trachea at the thoracic inlet with impingement on the trachea by a tortuous brachiocephalic artery. This patient underwent tracheal reconstruction and arterial relocation by our cardiothoracic surgeon [27] with good results. Prior to this incident, we had assumed that the airway obstruction in these patients was entirely due to the narrow upper airway passages, beginning at the nostril and continuing distally to the oropharyngeal and supra glottic structures [7], compounded by the short neck with limited head and neck motion [28]. However, following the above-mentioned patient's critical airway findings in 2015, we began evaluating our patients with Morquio A patients using CTA of the chest, and we reported our findings in a large cohort of patients [16, 25]. Our most recent study [9] examining anesthetic difficulties related to the tracheal narrowing in patients with Morquio A showed that as early as age 7 years, patients may have significant tracheal narrowing. This finding prompted us to lower the age to 7 years for initial CTA evaluation of the trachea. Although CTA provides the most detailed assessment of the trachea, vessels, bones, and other tissues, our radiologists also now specifically evaluate the trachea in these patients when visible on other imaging modalities, such as spine magnetic resonance imaging (MRI) or scoliosis radiographs (Fig. 14.5). Information about tracheal abnormalities is critical in preparing for the perioperative care of our patients with Morquio A [9].

The tracheal pathology of Morquio A patients has stirred a great deal of academic curiosity due to its unique anatomical appearance. While tracheal abnormalities may occur, albeit rarely, in other disorders (including other MPS patients [29]), significant tracheal pathology is consistently present in Morquio A [9, 16, 25]. These studies have reported that by age 15 years, patients with the severe phenotype of Morquio A have developed at least 50% tracheal narrowing at the thoracic inlet. This unusual abnormality of the trachea results from an imbalance between the longitudinal growth of the trachea and the abnormally short thoracic cage. This rather simple explanation is increasingly gaining popularity and will continue to do so due to lack of any other plausible explanation [9, 16].

To clarify, there are two distinct levels of airway obstruction in patients with Morquio A; "first level" due to narrowed air passages from the nares to the larynx compounded by the short neck with limited mobility; and "second level" due to tracheal narrowing, buckling, and twisting at the thoracic inlet [9].

Pediatric anesthesiologists may encounter patients with Morquio A more frequently than adult providers due to the multiple surgical procedures the patients undergo beginning at an early age [7]. However, patients with Morquio A are living longer with current modes of treatment and are expected to live into their 30s with the severe phenotype and up to their 70s with the mild phenotype [30, 31]. Therefore, both pediatric and adult anesthesia providers may encounter patients with Morquio A and need to be familiar with the anesthetic risks posed by their co-morbid conditions.

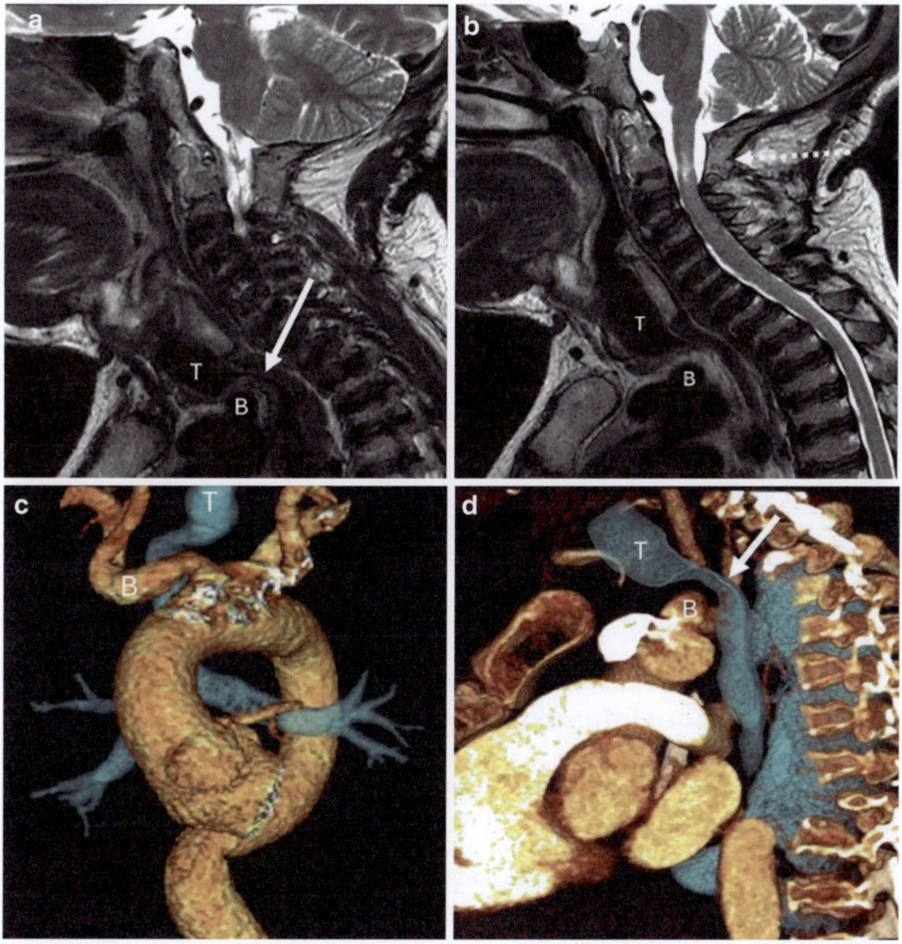

Fig. 14.5 28-year-old man with Morquio A syndrome. Two sagittal T2-weighted MR images of the cervical spine (**a, b**) show severe tracheal narrowing and buckling at the thoracic inlet (arrow), with the brachiocephalic artery (B) causing direct impingement on the trachea (T). This patient has bony occipital to C2 fusion (dashed arrow) with focal myelomalacia in the upper cervical spinal cord. Reconstructions from CT angiography of the chest from the patient in Fig. 14.4 performed 2 years earlier shows the tracheal, vascular, and bony anatomy in greater detail. Three-dimensional surface rendered image of the trachea and arteries (**c**) shows tortuosity of the aorta and great vessels (orange) and buckled, "worm-like" configuration of the trachea (teal) as it drapes over the brachiocephalic artery. A sagittal thick-slab reconstruction (**d**) shows similar features with bones and lung also included. Composite image from centerline reconstruction technique. (**e**) Narrow and tortuous trachea shown in two dimensions, allowing for virtual bronchoscopy (**f**, pink), displaying the slit-like narrowing of the trachea at the thoracic inlet as seen from just below the vocal cords (white arrow). Annotations for orientation from the 3D software include: F (foot), H (head), R (right), L (left), A (anterior), P (posterior)

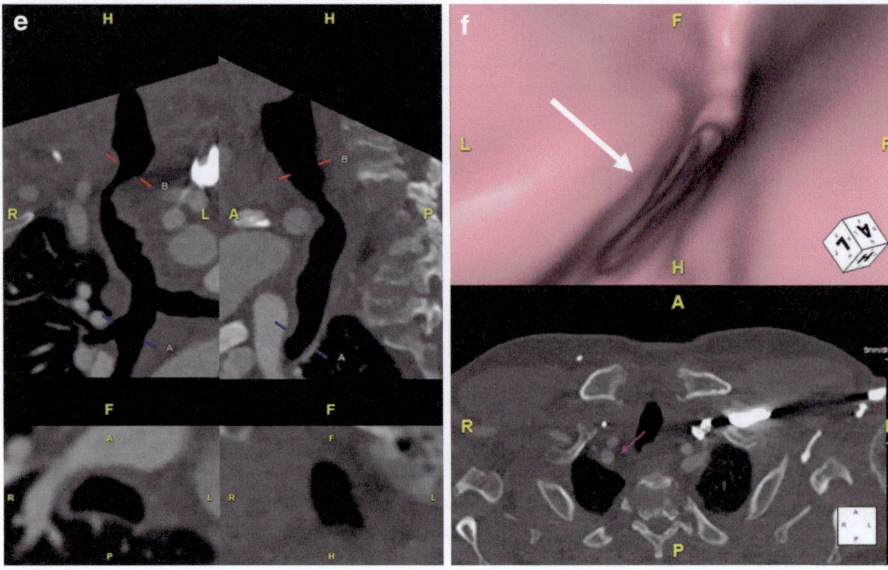

Fig. 14.5 (continued)

14.5 Spinal Cord Infarction in Morquio A

The spinal cord in patients with severe Morquio A is susceptible to injury with any procedure or maneuver that may compromise perfusion. Such situations include regional anesthetics that are neuraxial based, such as spinal or epidural anesthesia. There are enough cases reported (and unreported) that we consider spinal or epidural anesthesia in patients with Morquio A contraindicated [10, 11].

In one report [10], a 16-year-old patient with Morquio A had been anesthetized with a combination of general and epidural anesthesia for bilateral distal femoral osteotomies. Prior imaging studies of the spine had reported several areas of compromise of the spinal cord, including moderate stenosis of the spinal cord at C2–6 and T1–4. Mild kyphosis of the thoracolumbar junction and minimal CSF around the spinal cord at T11–L1 was also mentioned. Although not mentioned in the radiology report, an exaggeration of normal kyphosis was observed at T2–4. Flexion-extension maneuvers done during the MRI studies did not result in further compromise of the cord. For the anesthesia care, a lumbar epidural was placed at the L2–3 level with the patient awake, followed by induction of general anesthesia. Following an anesthetic of approximately 5 h duration, the patient emerged from the general anesthesia but was soon observed to have a loss of both sensory and motor function below the T4 level. Due to suspected epidural hematoma, an MRI was performed, which revealed a lesion interpreted as spinal cord infarction from T2–5 with adjacent spinal cord edema through T7. Thus, the patient remained paraplegic at the T4 level.

While this is the first reported case of an irreversible spinal cord infarction following an anesthetic that included an epidural in a patient with Morquio A, we are aware of several instances of spinal cord injury (presumably "infarction") in not only patients with Morquio A, but also in patients with other types of MPS. Not all patients who sustained such injury to the spinal cord had an epidural or spinal anesthetic. The exact mechanism of the injury is unclear, but given the devastating nature of this complication, we consider spinal and epidural anesthetics contraindicated in MPS patients. The "Wake Up Safe" anesthesia patient safety initiative by the Society of Pediatric Anesthesia has published a statement to that effect as well.[1]

We practice neuromonitoring of patients with skeletal dysplasia undergoing surgery and anesthesia who have a "spine at risk."[2] While nearly all patients with MPS would typically have a "spine at risk," many other forms of skeletal dysplasia would exhibit such characteristics as well [32–36]. At the time of this writing, we continue to serve as a resource to advocate for neuromonitoring of all patients who have a "spine at risk" and are undergoing surgery and/or anesthesia for longer than an hour. This prompts another question related to anesthesia and sedation needed for medical imaging studies that are often longer than an hour. For patients who are noted to have a "spine at risk," we would either abbreviate the medical imaging study to an hour (or less), or, as a second option, perform the imaging study in the lateral position [37].

14.6 Controversy Over Lumbar Plexus Block

The possible epidural spread of local anesthetic when a lumbar plexus block is performed poses concern in patients with a "spine at risk." Lumbar plexus in combination with a general anesthetic is most frequently performed for hip procedures in patients with skeletal dysplasia.

A lumbar plexus block (LPB), also known as a psoas compartment block, is a regional anesthetic technique that most commonly targets L2, L3, and L4 on the ipsilateral side. The procedure is performed to provide postoperative pain relief for surgical procedures involving the hip joint. The nerves targeted are the ilioinguinal, genitofemoral, lateral femoral cutaneous, femoral, and obturator nerves. When certain conditions are met (volume of the local anesthetic injected as well as the pressure of the injection), an LPB could result in the spread of the local anesthetic to the epidural space. It has been reported that bilateral spread can be seen with LPBs despite a unilateral injection up to 27% of the time [38]. Injection pressure greater than 20 psi has resulted in a 50% incidence of neuraxial spread [38]. Eight of 30 (26.7%) patients in one study had atypical and extended reduced sensory test scores that indicated local anesthetic spread to the epidural space or sacral plexus [39]. In another study, MRI demonstrated spread of local anesthetic spread to the epidural space in 5 of 17 (29.4%) healthy volunteers [40]. These numbers prove that epidural spread of local anesthetic is common in LPBs. Given the devastating nature of the

[1] https://wakeupsafe.org/wp-content/uploads/2018/09/MPS_statement.pdf

[2] Please refer to part III, Chap. 17. Anesthetic management of spine fusion

complications associated with an epidural anesthetic in patients with Morquio A, LPB in these patients raises a red flag. For this reason, we do not advocate the use of LPBs in patients with MPS.

14.7 Case Illustration

14.7.1 Patient 1

A 33-year-old female diagnosed with Morquio A syndrome was assessed first by our orthopedic surgeons for progressive difficulty with walking at age 15 years. Prior to coming to Nemours, she had been followed in the past by multiple physicians in the New York City area. Her medical history included a significant event at age 1, when she fell and stopped breathing. She was taken to the hospital, but her breathing had improved significantly by the time she arrived in the emergency room. The physicians who examined her did not find anything abnormal at that time, and she was sent home. However, her mother noticed some weakness in her walking ability, and she was advised to see a neurologist. The resulting work-up showed C1–2 instability, leading to a laminectomy/decompression surgery of the cervical spine, C1–2. At that time, she had not yet been diagnosed with Morquio syndrome. Up to that point, she had met all her developmental milestones and even cruised with a halo on. After her spinal surgery, she continued to walk, but her strength plateaued, and she could not walk long distances. At age 8 years, she underwent leg realignment surgery, which did not help or improve her walking. Subsequently, she seemed to get progressively weaker and had not walked significantly since then. She occasionally used a walker or crutches, but only for very short distances. Her leg deformities subsequently recurred.

Anesthesia Care At age 17, she had her first anesthetic at Nemours Children's Hospital, Delaware, for bilateral lower extremity osteotomies and a realignment procedure. A video laryngoscope was used to intubate the trachea, and, while the chart does not mention difficulty with mask ventilation, it is likely that there was some degree of upper airway obstruction during face mask ventilation—possibly requiring a two-person technique. At age 18, when undergoing a cervical spine fusion, she again was intubated using a video laryngoscope. Her pre-surgical X-rays of the cervical spine in flexion and extension show limited flexion with adequate extension (Fig. 14.6).

In her 33 years of life, she had undergone multiple surgical procedures, including cervical spine fusion, lower extremity osteotomies, and tracheal resection and reconstruction at age 23 years for significant tracheal narrowing. Patients with Morquio A were, at the time, beginning to be noted for their tracheal narrowing. Upon evaluation, her trachea was noted to be narrow at the thoracic inlet by 75% (when compared to the intrathoracic portion of her trachea). Prior to her surgery, she underwent a flexible bronchoscopic evaluation of her trachea in the operating room under minimal sedation. She had recovered well from her bronchoscopic evaluation

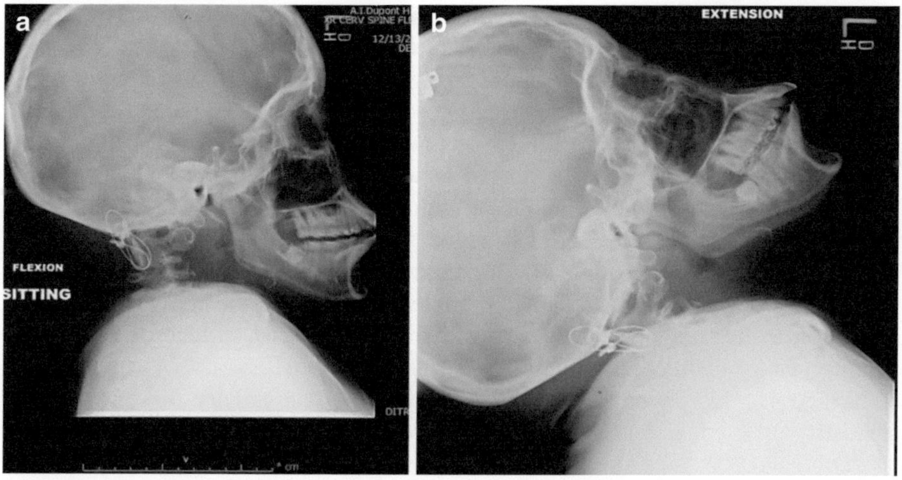

Fig. 14.6 Flexion (**a**) and extension (**b**) of cervical spine at age 18 years with evidence of prior cervical spine surgery. Difficulties including a very short neck and limited overall mobility are expected during airway management

and was discharged with her parents. However, later that night she developed respiratory difficulty and was rushed to the emergency room where she was treated with steroids and nebulized epinephrine. She improved quickly and was discharged home following plans for tracheal resection and reconstruction surgery, which she successfully underwent 2 months later. Currently, at age 33 years, she is being evaluated for hip replacement.

14.7.2 Patient 2

A 41-year-old female with a diagnosis of Morquio A had been under our care since age 34 years. She first came to Nemours to participate in a study related to the natural history of Morquio A. She had had several surgeries prior to coming to Nemours: these included bilateral tibial osteotomies at age 6.5, femoral and tibial osteotomies at age 11, a cervical fusion at age 20, and a right total hip replacement at age 25. These surgeries were performed elsewhere, and we had no information regarding airway management. She had other stigma of Morquio A, including corneal clouding, bilateral mixed hearing loss, mitral valve regurgitation, and moderate aortic stenosis. In addition, she had a history of mild obstructive sleep apnea and was followed by a pulmonologist in her community.

Her first surgical procedure at Nemours was a cochlear implant. As it was her first anesthetic under our care, we were uncertain about the difficulty regrading her airway. Her mouth opening was somewhat limited but adequate for using a video laryngoscope (Glidescope; Verathon, Bothell, WA). A CT angiogram performed to evaluate her trachea revealed a 50% narrowing of the trachea at the thoracic inlet.

During induction of anesthesia, ventilation with face mask was adequate with a two-person technique. However, upon attempting to intubate the trachea using the GlideScope, despite tongue retraction, we were unable to access her larynx, even though we had a partial but adequate view of the vocal cords. Prepared for this possibility, we had a flexible bronchoscope loaded with an endotracheal tube at the ready. Thus, her trachea was intubated using a hybrid technique where a video laryngoscope was used in combination with a flexible bronchoscope [41]. In this patient's case, the flexible bronchoscope was introduced nasally and rapidly advanced into the trachea, aided by the GlideScope view of the larynx. This technique worked well for her, and she did not experience any side effects other than a laceration to her tongue, which occurred when the dorsal surface of her tongue scraped against her teeth during tongue retraction.

At age 41, this patient is currently awaiting cardiac surgery for severe aortic stenosis.

References

1. Tomatsu S, Montaño AM, Oikawa H, Smith M, Barrera L, Chinen Y, et al. Mucopolysaccharidosis type IVA (Morquio A disease): clinical review and current treatment. Curr Pharm Biotechnol. 2011;12:931–45.
2. Muenzer J. Overview of the mucopolysaccharidoses. Rheumatology (Oxford). 2011;50(Suppl 5):v4–12.
3. Muhlebach MS, Wooten W, Muenzer J. Respiratory manifestations in mucopolysaccharidoses. Paediatr Respir Rev. 2011;12:133–8.
4. Hendriksz CJ, Harmatz P, Beck M, Jones S, Wood T, Lachman R, et al. Review of clinical presentation and diagnosis of mucopolysaccharidosis IVA. Mol Genet Metab. 2013;110:54–64.
5. Lee YH, Su CH, Lin CY, Lin HY, Lin SP, Chuang CK, et al. Endoscopic and image analysis of the airway in patients with mucopolysaccharidosis type IVA. J Pers Med. 2023;13:494.
6. Morimoto N, Kitamura M, Kosuga M, Okuyama T. CT and endoscopic evaluation of larynx and trachea in mucopolysaccharidoses. Mol Genet Metab. 2014;112:154–9.
7. Theroux MC, Nerker T, Ditro C, Mackenzie WG. Anesthetic care and perioperative complications of children with Morquio syndrome. Paediatr Anaesth. 2012;22:901–7.
8. Theroux MC, Lopez M, Olszewsky PJ, DiCindio S, Arai L, Ditro C, et al. Metatropic dysplasia—a skeletal dysplasia with challenging airway and other anesthetic concerns. Paediatr Anaesth. 2017;27:596–603.
9. Theroux MC, DiCindio S, Averill LW, Pizarro C, Oommen A, Bober MB, et al. Tracheal narrowing and its impact on anesthesia care in patients with Morquio A (mucopolysaccharidosis type IVA): an observational study. Anesth Analg. 2023;137:1075–83.
10. Drummond JC, Krane EJ, Tomatsu S, Theroux MC, Lee RR. Paraplegia after epidural-general anesthesia in a Morquio patient with moderate thoracic spinal stenosis. Can J Anaesth. 2015;62:45–9.
11. Tong CK, Chen JC, Cochrane DD. Spinal cord infarction remote from maximal compression in a patient with Morquio syndrome. J Neurosurg Pediatr. 2012;9:608–12.
12. Tomatsu S, Montaño AM, Nishioka T, Gutierrez MA, Peña OM, Tranda Firescu GG, et al. Mutation and polymorphism spectrum of the GALNS gene in mucopolysaccharidosis IVA (Morquio A). Hum Mutat. 2005;26:500–12.
13. Tomatsu S, Dieter T, Schwartz IV, Sarmient P, Giugliani R, Barrera LA, et al. Identification of a common mutation in mucopolysaccharidosis IVA: correlation among genotype, phenotype, and keratan sulfate. J Hum Genet. 2004;49:490–4.

14. Haskins ME. Animal models for mucopolysaccharidosis disorders and their clinical relevance. Acta Paediatr. 2007;96:56–62.
15. Harmatz PR, Mengel KE, Giugliani R, Valayannopoulos V, Lin SP, Parini R, et al. Longitudinal analysis of endurance and respiratory function from a natural history study of Morquio A syndrome. Mol Genet Metab. 2015;114:186–94.
16. Averill LW, Kecskemethy HH, Theroux MC, Mackenzie WG, Pizarro C, Bober MB, et al. Tracheal narrowing in children and adults with mucopolysaccharidosis type IVA: evaluation with computed tomography angiography. Pediatr Radiol. 2021;51:1202–13.
17. Kampmann C, Abu-Tair T, Gökce S, Lampe C, Reinke J, Mengel E, et al. Heart and cardiovascular involvement in patients with mucopolysaccharidosis type IVA (Morquio-A syndrome). PLoS One. 2016;11:e0162612.
18. Factor SM, Biempica L, Goldfischer S. Coronary intimal sclerosis in Morquio's syndrome. Virchows Arch A Pathol Anat Histol. 1978;379:1–10.
19. Käsmann-Kellner B, Weindler J, Pfau B, Ruprecht KW. Ocular changes in mucopolysaccharidosis IV A (Morquio A syndrome) and long-term results of perforating keratoplasty. Ophthalmologica. 1999;213:200–5.
20. Cahane M, Treister G, Abraham FA, Melamed S. Glaucoma in siblings with Morquio syndrome. Br J Ophthalmol. 1990;74:382–3.
21. Nagao K, Walter C, Parkes WJ, Teixido M, Theroux MC, Szymkowski S, et al. Cochlear implantation in a patient with mucopolysaccharidosis IVA. SAGE Open Med Case Rep. 2019;7:2050313X19873791.
22. Riedner ED, Levin LS. Hearing patterns in Morquio's syndrome (mucopolysaccharidosis IV). Arch Otolaryngol. 1977;103:518–20.
23. Yamauchi K, Hirano D, Wada M, Ida H. Pectus carinatum as the key to early diagnosis of Morquio A syndrome: a case report. J Med Case Rep. 2021;15:150.
24. Gardner DG. The dental manifestations of the Morquio syndrome (mucopolysaccharidosis type IV). A diagnostic aid. Am J Dis Child. 1975;129:1445–8.
25. Tomatsu S, Averill LW, Sawamoto K, Mackenzie WG, Bober MB, Pizarro C, et al. Obstructive airway in Morquio A syndrome, the past, the present and the future. Mol Genet Metab. 2016;117:150–6.
26. Doherty C, Averill LW, Theroux M, Mackenzie WG, Pizarro C, Mason RW, et al. Natural history of Morquio A patient with tracheal obstruction from birth to death. Mol Genet Metab Rep. 2017;14:59–67.
27. Pizarro C, Davies RR, Theroux M, Spurrier EA, Averill LW, Tomatsu S. Surgical reconstruction for severe tracheal obstruction in Morquio A syndrome. Ann Thorac Surg. 2016;102:e329–31.
28. Dede O, Thacker MM, Rogers KJ, Oto M, Belthur MV, Baratela W, et al. Upper cervical fusion in children with Morquio syndrome: intermediate to long-term results. J Bone Joint Surg Am. 2013;95:1228–34.
29. Kampmann C, Wiethoff CM, Huth RG, Staatz G, Mengel E, Beck M, et al. Management of life-threatening tracheal stenosis and tracheomalacia in patients with mucopolysaccharidoses. JIMD Rep. 2017;33:33–9.
30. Lavery C, Hendriksz C. Mortality in patients with Morquio syndrome A. JIMD Rep. 2015;15:59–66.
31. Montaño AM, Tomatsu S, Gottesman GS, Smith M, Orii T. International Morquio A Registry: clinical manifestation and natural course of Morquio A disease. J Inherit Metab Dis. 2007;30:165–74.
32. Pruszczynski B, Mackenzie WG, Rogers K, White KK. Spinal cord injury after extremity surgery in children with thoracic kyphosis. Clin Orthop Relat Res. 2015;473:3315–20.
33. Sahin MC, Ipek EM, Zinnuroglu M, Borcek AO. How did intraoperative neuromonitorization prevent tetraplegia? Childs Nerv Syst. 2023;39:1323–8.
34. Solanki GA, Martin KW, Theroux MC, Lampe C, White KK, Shediac R, et al. Spinal involvement in mucopolysaccharidosis IVA (Morquio-Brailsford or Morquio A syndrome): presentation, diagnosis and management. J Inherit Metab Dis. 2013;36:339–55.

35. Charrow J, Alden TD, Breathnach CA, Frawley GP, Hendriksz CJ, Link B, et al. Diagnostic evaluation, monitoring, and perioperative management of spinal cord compression in patients with Morquio syndrome. Mol Genet Metab. 2015;114:11–8.
36. White KK, Bober MB, Cho TJ, Goldberg MJ, Hoover-Fong J, Irving M, et al. Skeletal Dysplasia Management Consortium. Best practice guidelines for management of spinal disorders in skeletal dysplasia. Orphanet J Rare Dis. 2020;15:161.
37. Brothers J, DiCindio S, Oommen A, Thacker M, Mackenzie S, Campbell J, Mackenzie W, Theroux M. Skeletal dysplasia and recommendations for IONM changes. Society for Pediatric Anesthesia Meeting, Austin, Texas April 2023.
38. Gadsden JC, Lindenmuth DM, Hadzic A, Xu D, Somasundarum L, Flisinski KA. Lumbar plexus block using high-pressure injection leads to contralateral and epidural spread. Anesthesiology. 2008;109:683–8.
39. Sauter AR, Ullensvang K, Niemi G, Lorentzen HT, Bendtsen TF, Børglum J, et al. The Shamrock lumbar plexus block: a dose-finding study. Eur J Anaesthesiol. 2015;32:764–70.
40. Bendtsen TF, Pedersen EM, Haroutounian S, Søballe K, Moriggl B, Nikolajsen L, et al. The suprasacral parallel shift vs lumbar plexus blockade with ultrasound guidance in healthy volunteers—a randomised controlled trial. Anaesthesia. 2014;69:1227–40.
41. Doyle DJ. GlideScope-assisted fiberoptic intubation: a new airway teaching method. Anesthesiology. 2004;101:1252.

Mucopolysaccharidosis Types I, II, III, V, VI and VII

15

Phoebe Fisher

15.1 Introduction

Mucopolysaccharidosis (MPS) refers to a group of disorders with the inability to breakdown glycosaminoglycans (GAGs) in lysosomes. Glycosaminoglycans are negatively charged polysaccharide compounds ubiquitous in human tissue. Their functions are diverse, from cell hydration and structural scaffolding to a complex role in cell signaling. Enzyme deficiencies in patients with MPS prevent the proper metabolism of GAGs. This causes intermediates to accumulate in lysosomes, leading to cellular disfunction and cell death. Glycosaminoglycans and their secondary storage metabolites also cause oxidative stress through activation of the inflammatory cascade [1]. The specific enzyme defect dictates which tissues have the greatest accumulation of GAGs, which in turn dictates the syndrome characteristics and disease progression.

In 1917, Dr. Charles Hunter first described a constellation of characteristics in two brothers that was to be known as Hunter syndrome, later classified as MPS II. In 1919, Gertrude Hurler described another pair of brothers with similar characteristics except they had a gibbus deformity and, of note, corneal clouding. Up until 1968, these were thought to be the same disorder. In 1978, Hunter and Hurler syndrome were identified as two distinct syndromes in a paper by A.E Lorincz [2, 3]. As more has been learned about this group of disorders, the classification of each has shuffled. Presently there are seven distinct types of MPS, each defined by the deficiency of a particular enzyme or enzymes that cause deposition of GAGs in different tissues.

P. Fisher (✉)
Anesthesiology and Pediatrics, Sidney Kimmel Medical College, Thomas Jefferson University, Philadelphia, Pennsylvania, USA

Department of Anesthesiology and Perioperative Medicine, Nemours Children's Hospital, Wilmington, Delaware, USA
e-mail: pfisher@nemours.org

M. C. Theroux (ed.), *Anesthesia Care of Patients with Skeletal Dysplasia*,
https://doi.org/10.1007/978-3-031-92501-6_15

MPS I has been divided into three subtypes: **Hurler**, **Hurler-Scheie**, and **Scheie**, defined by the age of onset and the severity of the clinical course. MPS II, **Hunter** syndrome, was once considered two distinct syndromes. It is now understood to be one syndrome with a spectrum of severity. The most severe course has features similar to, though not as debilitating as, Hurler. The mildest cases may not be evident until the second decade of life. MPS III, **Sanfilippo syndrome**, has four subtypes, each with a different specific enzyme defect. Sanfilippo syndrome is unique amongst the other MPSs as it is characterized primarily with neurodegeneration, causing progressive dementia and aggression; other physical characteristics common to MPS are usually only mildly expressed in these patients. MPS IV, **Morquio** syndrome, has two subtypes: A, more severe, and B, milder. Morquio A is fraught with extreme and progressive spine changes, resulting in spinal cord and nerve root compression and a tortuous trachea. A separate chapter in this book is dedicated to both Morquio A and B[1]. MPS V is now understood to be a form of MPS I and is no longer a separate classification. MPS VI, **Maroteaux-Lamy** syndrome is similar to the more severe form of MPS I without the cognitive decline. MPS VII, **Sly** syndrome, often presents as hydrops fetalis. Those with MPS VII who do survive may live into their teens with vision loss, joint problems, and repeated bouts of pneumonia. MPS VIII is an obsolete classification. It represented a single individual with a mixed MPS III and IV presentation. MPS IX, **Natowicz** syndrome is very rare, with only four known cases. The syndrome presents with periarticular soft-tissue masses. Other than mild short stature, these patients present almost identical to patients with juvenile idiopathic arthritis and have little else in common with the other MPS syndromes [4–6].

15.2 Genetics

There are seven types of MPS. Types I, III, IV, VI, VII, and IX are all inherited in an autosomal-recessive pattern. MPS II is x-linked with symptoms rarely seen in girls. Related genes as commonly reported are type I, *IDUA*; II, *IDS*; III, *SGHS* (MPS3A); IV, *GALNS* (MPS4A) and *GLB1* (MPS4B); VI, *ARSB*; VII, *GUSB*; and IX, *ARSK* [7] (Table 15.1). There is a common founder effect such that there is a particularly high incidence of MPS in Saudi Arabia with 16.9 cases in 100,000 births. The country with the next highest incidence is Portugal at 4.8 per 100,000 births. The United States has an incidence of 1.2 per 100,000 births. The country with the highest prevalence is Brazil with 1652 cases from 1982 to 2019, followed by Germany and Japan. Prenatal diagnosis and preimplantation genetic testing will influence the ongoing prevalence of these diseases [4].

[1] Please refer to Chap. 14: Morquio A (MPS IVA) Syndrome.

Table 15.1 Different types of MPS with the associated enzyme deficiency, resultant glycosaminoglycan accumulated, incidence rate, and mode of inheritance

			Enzyme defect [4, 7]	GAG accumulation [4, 7]	Incidence	Genetics
MPS I	Hurler		α-L-Iduronidase	Heparin Sulfate, Dermatan Sulfate	1:100,000 [8]	Autosomal recessive
	Hurler-Scheie			Dermatan Sulfate	1:115,000 [8]	Autosomal recessive
	Scheie			Dermatan Sulfate	1:500,000 [8]	Autosomal recessive
MPS II	Hunter	A	Iduronidase-2-sulfatase	Heparin Sulfate, Dermatan Sulfate	1:100,000–1:150,000 males [8]	X-linked recessive
		B	Iduronidase-2-sulfatase	Heparin Sulfate, Dermatan Sulfate		
MPS III	Sanfilippo	A	Heparin-N-sulfatase	Heparin Sulfate	1:50,000–1:250,000 [9]	Autosomal recessive
		B	α-N-acetylglucosaminidase	Heparin Sulfate		
		C	α-glucosaminidase acetyltransferase	Heparin Sulfate	1:1,500,000–1:1,000,000 [9]	
		D	N-acetylglucosamine-6-sulfatase	Heparin Sulfate		
MPS IV	Morquio	A	N-acetylglucosamine-6-sulfate sulfatase	Chondroitin 6-Sulfate, Keratan Sulfate	1:150,000–1:470,000 [10]	Autosomal recessive
		B	β-galactosidase	Keratan Sulfate	1:250,000–1:1,000,000 [11]	Autosomal recessive
MPS VI	Maroteaux-Lamy		N-acetylglucosamine-4-sulfatase	Dermatan Sulfate, Chondroitin 6-Sulfate	1:250,000–600,000 [12]	Autosomal recessive
MPS VII	Sly		β-glucuronidase	Heparin Sulfate Dermatan Sulfate, Chondroitin 6-Sulfate, Chondroitin 4-Sulfate	1:300,000–1:2,000,000 [13]	Autosomal recessive
MPS IX	Natowicz		Hyaluronidase	Hyaluronan	Four known cases [4]	Autosomal recessive

15.3 Phenotype

Mucopolysaccharidosis is a group of hereditary disorders that have in common a deficiency in one of the enzymes that degrade GAGs. The partially degraded mucopolysaccharides accumulate in connective tissues and can affect all regions of the body, including the bones, brain, liver, blood vessels, skin, cartilage, airways, heart valves, and corneas [14]. There are variations in presentation both within and between the various MPS types.

There is significant respiratory system compromise, which includes both an obstructive and a restrictive pattern in patients with MPS, except MPS IX. Skeletally, patients with MPS have a short neck, a restricted temporomandibular joint motion, and some degree of micrognathia. There is progressive accumulation of GAG in the tissues of the upper airway, tongue, adenoids, tonsils, pharyngeal wall, and other soft tissues, including nasal turbinate. This leads to upper airway obstruction with sleep apnea, which at times is severe enough to require tracheostomies. Thoracic cage and spine abnormalities contribute to a restrictive pattern. Hepatosplenomegaly limits diaphragmatic excursion, adding to the restrictive pattern. A recent case study of 74 patients with MPS from Ankara, Turkey, showed adenoid hypertrophy in 65%, tonsillar hypertrophy in 54%, laryngomalacia in 28%, subglottic stenosis in 7%, and tracheomalacia in 42%. Additionally, they found that 81% were noted to be a difficult intubation [15]. Another study compared the rate of complications after adenotonsillectomy in 50 patients with MPS to other children. They found a 15% rate of respiratory complications in children with MPS versus 2.4% in all others. Complications included respiratory compromise, postoperative hemorrhage, increased length of stay, and higher costs [16].

Cardiac disease is common, with an early onset in all types of MPS, but particularly, in MPS I, II, and VI. Dermatan-sulfated GAGs are a normal and prominent component of cardiac valves. The enzyme deficiencies in MPS I, II, and VI lead to the accumulation of Dermatan-sulfated GAGs. Thus, cardiac valve thickening and dysfunction is seen in 80% of Hurler, 57% of Hunter, and 100% of Maroteaux-Lamy patients [17]. Importantly, cardiovascular disease is often silent in these patients. Murmurs may be absent in mild-to-moderate valve disease or difficult to detect due to body habitus and developmental limitations. In one report, only 6 of 26 patients with disease documented by cardiac echo had signs and symptoms consistent with their degree of disease [17]. Most valvular thickening leads to regurgitation rather than stenosis. The mitral valve is typically the most affected, followed by the aortic valve, while the least affected are the right-sided valves. Case studies have shown the most common progression of cardiac disease to be valvular disease leading to left ventricular hypertrophy and diastolic dysfunction in the earlier stages, with ventricular dilation and

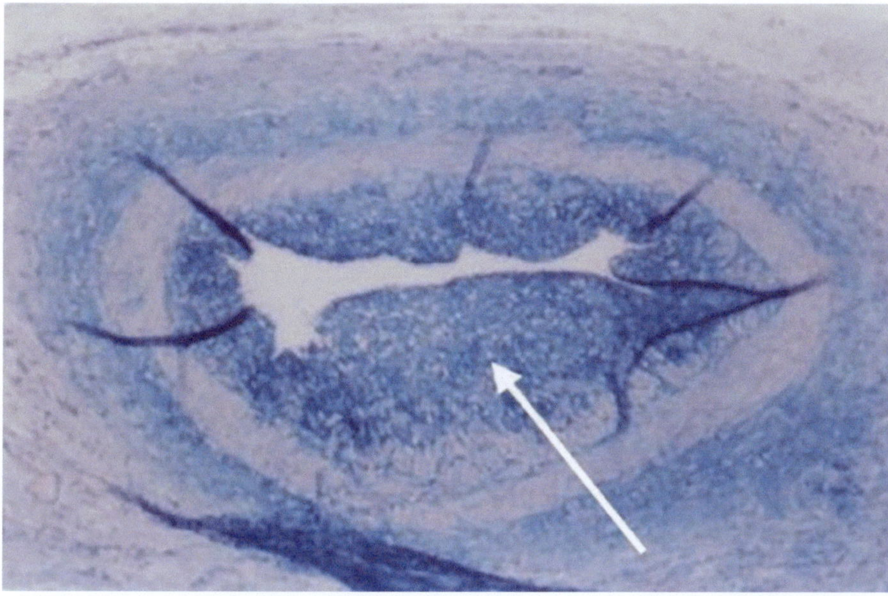

Fig. 15.1 Epicardial right coronary artery from an untreated 1-year-old with MPS I showing diffuse myo-intimal proliferation [17, 18]

systolic dysfunction occurring later. There is also diffuse infiltration of GAGs in coronary arterioles causing central narrowing of the lumens. GAG deposits may also interfere with the conduction system (Fig. 15.1). There have been case reports of complete atrioventricular block, requiring pacing or leading to sudden death [17].

The skeletal abnormalities of the MPS disorders are relatively uniform and can be described as dysostosis multiplex [19]. Dysostosis multiplex includes macrocephaly with a thick skull, oar shaped ribs, atlantoaxial instability, cervical stenosis, hook-shaped vertebral bodies with anterior beaking or wedging causing a thoracolumbar gibbus deformity, coxa valga, hip dysplasia, and abnormalities of the tubular bones with metaphyseal and epiphyseal dysplasia.

Other common characteristics of the phenotype include vision loss, primarily from corneal clouding (not in MPS II and III), but also from retinopathy and involvement of the sclera, trabecular meshwork, and optic nerve [20] (Table 15.2). Hearing difficulties, poor dentition including dental carries, carpel tunnel syndrome, and umbilical and inguinal hernias are also common.

Table 15.2 Different types of MPS with severity of phenotype (based on sub types), natural history, life expectancy and the treatment modalities are given

			Natural history			Corneal clouding	Treatment [21]
			Severity	Neurocognition	Life expectancy		
MPS I	Hurler		Severe	Problems appear in the second year with slowing then halting of milestones by age 4 followed by decline	Early childhood	Yes	ERT and HSCT
	Hurler-Scheie		Intermediate	This represents more of a spectrum, with mild impairment to normal intelligence. Somatic disease may not correlate	Late teens to early 20s		ERT with optional HSCT
	Scheie		Mild		Before 30		
MPS II	Hunter	A	Severe	Presents just bit later than MPS IH, age 2–4 with a more variable course, may plateau for years before decline	Second decade	No	ERT (HSCT results equivocal)
		B	Mild	Neurocognitive difficulties without regression	Fifth or sixth decade		
MPS III	Sanfilippo	A	Variable	More severe forms are near normal until age 2, then regress until age 3–4 and fully dependent by teenage years. The more attenuated forms may live to early adulthood. The hallmark is severe behavioral problems with hyperactivity and aggression.	Second decade. The more attenuated forms may live to early adulthood	No	Supportive
		B	Variable				
		C	Variable				
		D	Variable				
MPS IV	Morquio	A	Variable	Normal, but some may have anxiety and depression	Late childhood to early teens	Yes	ERT with optional HSCT
		B	Mild		Adulthood, but reduced		ERT
MPS VI	Maroteaux-Lamy		Variable	Normal or near normal, attention problems. Severe disease may cause cognitive decline	Second decade in the severe forms	Yes	ERT with optional HSCT
MPS VII	Sly syndrome		Variable	Very rare, appears to be a range of intellectual disability from mild to severe	Variable	Yes	ERT with optional HSCT
MPS IX	Natowicz		Mild			Unknown	Supportive

15.4 MPS I

Hurler, Hurler-Scheie, and Scheie syndromes result from variants in the α-L-iduronidase (*IDUA*) gene. There are 200 known gene variants leading to MPS I. Hurler, the most severe, is phenotypically distinct from Hurler-Scheie and Scheie but, so far, not genotypically distinct [22]. Hurler-Scheie and Scheie have a broader and less delineated spectrum of presentation and are spared most of the intellectual decline. A patient with Hurler often appears normal at birth other than a gibbus deformity of the spine that may go unnoticed. Their initial contact with medical care is often prior to the MPS diagnosis for inguinal or umbilical hernia repairs or for adenotonsillar hypertrophy. The sooner the diagnosis, the sooner treatment can be started, which is particularly important in severe MPS I, Hurler. With neonatal screening newly available, efforts are being made to link genotype and phenotype to better define prognosis and treatment [22]. The natural history of Hurler is death by cardiorespiratory failure in the first decade of life. Hurler-Scheie and Scheie usually present between ages 3 and 10 years, and the natural history is survival to the second or third decade before also succumbing to cardiorespiratory failure [23].

15.5 MPS II

Hunter syndrome has a severe form and an attenuated form. MPS IIA, the severe form, presents a lot like Hurler, but without the corneal clouding. MPS IIB, the attenuated form, has milder somatic changes, clear cornea, and normal intelligence and might not be diagnosed until adolescence or early adulthood.

15.6 MPS III

Sanfilippo syndrome has milder somatic symptoms compared with the other MPS types but stands out for its central nervous system involvement. Early in life, the patients appear normal, but by age 1–3 years, they start to lose cognitive milestones and often develop aggression. By their teens, they can regress to a vegetative state [24].

15.7 MPS IV

Morquio syndrome is in on the spectrum with the other MPS presentations, including a short stature, but with a markedly short trunk and hyper-mobile joints. This leads to some complications that are unique to the Morquio patients as further described in the Morquio syndrome chapter in this book[2].

[2] Please refer Footnote 1.

15.8 MPS VI

Maroteaux-Lamy syndrome, like the other MPS disorders, has a spectrum of presentation. Generally, intelligence is normal or near normal with attention deficit and related concerns. More severe cases may have greater cognitive involvement. Patients have a spectrum of the same cohort of physical symptoms and often present first with corneal clouding, adenotonsillar hypertrophy, hernias, or carpel tunnel.

15.9 MPS VII

Sly syndrome is an ultra-rare disorder. Most patients may die in utero or in early infancy. At least 40% have non-immune hydrops fetalis. The occurrence of hydrops fetalis does not predict the course of the disease. For those that survive, there is a wide spectrum of disease, like in MPS I and MPS II, with similar presentation, including dysostosis multiplex, coarse facial features, developmental delay, hepatosplenomegaly, hernias, and progressive respiratory and cardiac involvement [13].

15.10 MPS IX

Natowicz syndrome is the rarest of the lysosomal storage disorders, with only four known cases, three of which are siblings of consanguineous parents [5]. In the siblings, the phenotype was limited to joint involvement, presenting similarly to juvenile rheumatoid arthritis. There is no anesthetic information available for these patients.

15.11 Diagnosis and Treatment

Historically, patients with MPS were diagnosed only once a constellation of physical attributes or cognitive decline were noted and a diagnostic connection made. As of 2020, all but eight states have MPS I in their mandatory newborn screening. Two states, Illinois and Missouri, have both MPS I and II [25]. Early treatment is critical to limit the long-term morbidity of these conditions.

Enzyme replacement therapy (ERT) has been developed for MPS I, II, IV, VI, and VII. Studies have shown that ERT may improve walking ability, endurance, and pulmonary function; however, in tissues with less perfusion, such as joints, cornea, and cardiac valves, the disease is slowed at best. Enzyme replacement therapy needs to be given intravenously once a week for life. The enzyme is a large molecule that cannot cross the blood–brain barrier; thus, it cannot help with the cognitive decline that is so devastating in the severe forms MPS I and II. Allergies, including anaphylaxis, to the enzyme may develop, reducing efficacy over time [26].

Hematopoietic stem cell transplant (HSCT) with stem cells from umbilical cord blood, bone marrow, or peripheral blood can improve symptoms of MPS and has been found to be effective in MPS I, II, IVA, VI, and VII. Tissues with greater perfusion show improvement following HSCT, leading to better joint mobility, vision, and hearing, and to a reduction in coarse facial features, upper airway obstruction, and hepatosplenomegaly. The disease continues to progress in tissues with less perfusion: bone, cornea, and cardiac valves. For MPS II and VII, HSCT has so far not been found to reverse any existing cognitive decline [21]. There is a small case series out of Australia and New Zealand showing cognitive stabilization in MPS II when HSCT was done before significant cognitive decline [27]. Hematopoietic stem cell transplant comes with significant risk. Prior to 2000, the mortality rate was up to 27%. This has improved significantly with better conditioning regimes [21]. Nonetheless, there are still significant side effects to HSCT that need to be weighed against ERT. The improvement in neurocognitive outcome in MPS IH and IIA makes HSCT in conjunction with a period of ERT presently the standard of care [21]. The same risk benefit analyses for IH-S and is not so clear. With MPS VI, HSCT is only considered when ERT has failed. There are a few case reports on the benefits of allogeneic HSCT in MPS VII, where it appears to reduce MPS symptoms and stop neurocognitive decline [28]. Research is now focusing on gene therapy using a viral vector. This shows great promise but still has many hurdles, both technical and financial, to overcome [29].

Understanding the fundamental aspects of available treatments is important for the effective anesthetic management of these patients. One must be aware that there are both visible and hidden processes in these diseases. For example, a patient may exhibit an improved range of motion and stamina with treatment while airway complications may only show partial improvement or just a plateau. Additionally, cardiac disease tends to progress over time and needs to be assessed thoroughly.

15.12 Nemours Children's Health Experience

The Nemours Children's Health database has more than 250 patients with MPS, including 34 with MPS I, 21 with MPS II, 9 with MPS III, and 17 with MPS VI (Table 15.3). All of these patients have had or will require surgery at some point. The spine, both at the cervical and the thoracolumbar levels, is at a high risk in these patients. Regarding management of anesthesia, airway management is always a primary concern. Face mask ventilation in young patients is generally easy or involves mild difficulty, though MPS II stands out as more challenging. Intubation should be done with a hyper-angulated video laryngoscope due to both the difficulty of the airway and the risk of cervical spine instability. As patients age, airway management becomes progressively more difficult, and greater difficulty should be anticipated. Difficult intubation tools including video laryngoscopes, flexible bronchoscopes, and laryngeal mask airways (LMA), should be readily available.

Table 15.3 Data from Nemours: Number (%) of patients in each MPS type, 'spine at risk' (kyphosis >20°) or spinal stenosis or instability, types of surgical procedures, and difficulty with airway management

		MPS I	MPS II	MPS III	MPS VI	MPS VII & IX
Number of patients in the Nemours skeletal dysplasia program		34	21	19	17	No patients
Thoracolumbar kyphosis > 20°		85%	37%	0%	73%	
Cervical stenosis with or without instability		32%	25%	0%	80%	
ENT procedure myringotomy, adenoidectomy and/or tonsillectomy		88%	90%	37%	94%	
Orthopedic lower extremity surgery in patients > 3 years old		71%	10%	0%	50%	
Mask induction	Easy mask induction with or without oral airway < 10 years	87%	62%	100%	92%	
	Progression of difficulty with mask ventilation for patients by 10 years	21%	20%	0%	28%	
Laryngoscopy and intubation	Easy with a hyper-angulated laryngoscope used for patients < 10 years	100%	56%	100%	91%	
	Progression of difficulty with intubation for patients by 10 years	36%	64%	0%	33%	
	Difficult or failed when a (conventional) miller or mac blade was attempted	50%	*100%	0%	83%	
	Required flexible bronchoscope-aided intubation by 7 years	20%	64%	0%	12%	

Note that in patients with MPS II, direct laryngoscopy was successful only in patients younger than 4 years. There was 100% failure rate in patients older than 4 years

15.13 Anesthesia Care

Anesthesia care starts with a careful review of the patient's history. MPS I, II, and VI are all at significant risk for anesthesia and surgery. There is limited information available related to the anesthesia management of MPS III in the literature. In a consensus guideline for MPS III published in 2022 by Muschol et al., a retrospective analysis of 126 procedures performed in 37 patients showed complications listed as bradycardia or tachycardia (2.4%), respiratory insufficiency (1.6%), hypoxemia (1.6%), and atelectasis (1.6%). They recommended that sedation and anesthesia events should always be conducted in the hospital setting where skilled anesthesia providers are available to manage airway emergencies [9]. There is no literature related to anesthesia in the four known patients with MPS IX.

15.13.1 Airway

Information from a prior induction and intubation is very helpful, but expect to have greater difficulty with increasing age. Our experience indicates that mask induction may be safe, particularly with younger patients, but caution needs to be exercised and a backup plan should be in place. If proceeding with an inhalational induction, attempting the first intravenous line early during the induction process is advisable. Most of these patients may be intubated with a hyper-angulated video laryngoscope, but an LMA and flexible bronchoscope (FB) should be readily available. When there is significant macroglossia, limited neck mobility, and limited mouth opening, a Magill forceps can be used to retract the tongue out of the mouth to create more space in the oropharynx either for a video laryngoscope or an FB.

15.13.2 Cardiopulmonary

Many patients with MPS will be using non-invasive ventilation devices, such as continuous positive airway pressure (CPAP) or bilevel positive airway pressure (BiPAP). Patients should be advised to bring their own masks or nasal prongs for their anticipated use in the postoperative period. Older patients are often put on antihypertensives, such as ACE inhibitors or ARBs for hypertension and left ventricular hypertrophy. These need to be held for 24 h (as is common practice for all patients). As patients get older, a cardiac evaluation becomes increasingly important, particularly for those at the highest risk of valvular disease—MPS I, II, IV, and VI. For MPS IV cardiac involvement and risk see Chap. 14.

15.13.3 Spine Precautions

The cervical and the thoracolumbar regions both need to be carefully considered. There is a high incidence of cervical stenosis with or without cervical instability in these patients. A shoulder roll needs to be used to align the external auditory meatus with the clavicle to keep the neck in a neutral position,[3] especially considering the presence of macrocephaly. Flexion and extension should be avoided unless proven safe by imaging. A hyper-angulated video laryngoscope is likely to be the safest choice for intubating the trachea. The thoracolumbar spine is also a significant source of risk. In our experience, 90% of patients with MPS I, 30% of MPS II, and 73% of MPS VI have a gibbus with a wedge vertebral deformity resulting in a greater than 20° kyphotic curve. This puts them at a significant risk for a spinal cord injury from just laying supine under anesthesia. There is a case report of two patients with a kyphosis of 76° and 80° who woke up with a spinal cord injury following procedures for hip dysplasia [30]. Our practice is that when the patient has a kyphosis greater than 20°, neuromonitoring needs to be employed to continuously assess the spinal cord, if an anesthetic exceeds

[3] Please refer to Chap. 16: Anesthetic Management of Spine Fusion.

1 h. When neuromonitoring is not possible, such as in the MRI suite, the procedure must be done in the lateral position. In such high-risk patients, it will be prudent to maintain an optimum mean arterial pressure, which should be roughly 20% above baseline. Of importance, the rescue position in the event of a loss of evoked potentials is to turn patient to the lateral position to take the pressure off the spinal cord.

15.13.4 Pain Control

These patients undergo many surgical procedures. Both lower extremities being operated during the same anesthetic is not unusual. While a spinal or an epidural is often employed in the non-skeletal dysplasia population, such neuraxial anesthetic poses significant risk to the patients with MPS. The Society of Pediatric Anesthesia highlights patients with MPS in their WakeUp Safe initiative, emphasizing the need to avoid neuraxial anesthesia as well as the need for neuromonitoring when there is a kyphotic curve present [31]. Spinal cord injury during an anesthetic had been reported in these patients during surgical procedures regardless of the surgical site. Such spinal cord injury has occurred both with and without a neuraxial anesthetic being used [32, 33]. In lieu of neuraxial techniques, peripheral nerve blocks can work very well. Due to the need for neuromonitoring, peripheral blocks need to be placed at the end of the procedure, just prior to emergence. This timing can be challenging due to the presence of surgical dressings and casts, requiring a creative and collaborative approach to succeed. Our practice at Nemours is that hip dysplasia procedures will usually receive a facia iliacus block with or without catheters. Though a lumbar plexus block could give very good coverage, due to the possibility of epidural spread of local anesthetic [34], we avoid it in patients with MPS. For more detailed explanation of concern related to lumbar plexus block, please refer to Chap. 14 on MPS IVA. For the various lower extremity epiphysiodesis procedures, femoral, adductor canal, and sciatic blocks can be used as appropriate.

15.14 Case Illustration

At age 5 years, a child with Maroteaux-Lamy syndrome (MPS VI) was referred to us from another university hospital for further care. He was diagnosed with MPS VI just before his second birthday due to coarse facial features and pectus excavatum, and he was started on galsulfase (Naglazyme; Biomarin, Novato, CA, USA) IV weekly. By age 2 years, he had undergone a posterior fossa decompression and C1 laminectomy, bone marrow transplant, adenotonsillectomy, and bilateral carpel tunnel releases at his home institution. His airway was managed by an endotracheal tube (ETT) through an LMA for one procedure and a GlideScope (Verathon, Bothell, WA, USA) for another. He came to Nemours when he developed progressive bilateral hip pain with walking due to hip dysplasia and coxa valga. Of note, imaging studies of the spine had shown a 40° thoracic kyphosis with anterior wedging of L2 and retrolisthesis of L2 on L3. His scheduled procedure at Nemours was bilateral hip arthrograms, proximal femoral osteotomies, and Dega osteotomies. He had an uneventful induction of anesthesia, during which he was positioned supine with a shoulder roll. Ventilation with face mask was possible with a "two-hand technique," following which he was intubated with a hyper-angulated video laryngoscope (GlideScope Spectrum; Verathon) without difficulty.

Following intubation, neuromonitoring leads were placed. Establishing baselines was prolonged due for the need to troubleshoot for unexpected findings. The upper-extremity signals were intact. Motor evoked potentials for the iliopsoas were able to be monitored; rectus muscle signals were labile with variable amplitude, but no responses from the tibialis anterior nor the abductor hallucis muscle groups were recordable. This was not consistent with the preoperative history. The decision was made to abort the procedure, which had been started during the troubleshooting period. He was promptly repositioned on his side to take pressure off his thoracolumbar kyphosis, allowing the skin incisions of one hip and then the other to be closed. The mean arterial pressure was increased using vasopressors to optimize spinal cord perfusion. The signals did not improve, and the patient was allowed to emerge sufficiently from the anesthetic to check lower extremity function. He moved toes bilaterally, and then was re-anesthetized and kept in a lateral position for an urgent MRI. The MRI showed his focal kyphosis centered at L2 and a normal appearance of the spinal cord with no evidence of compression or restricted diffusion (Fig. 15.2).

It was clear that his compromised spinal cord needed to be decompressed and his spine fused in situ prior to addressing any other surgical needs. He was brought back to the OR 12 days later for a posterior spinal fusion from T10–L4. He was kept in a left-lateral position for induction, intubation with a GlideScope, and the establishment

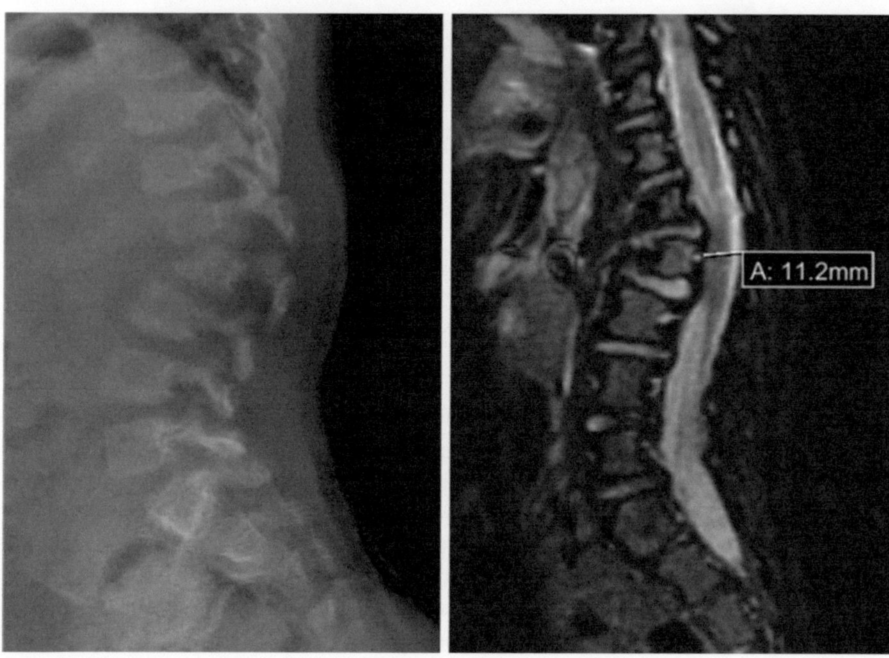

Fig. 15.2 View of the gibbous deformity on x-ray with an MRI image of the same area showing the spinal cord being tented over the marked prominence of the gibbus within the spinal canal

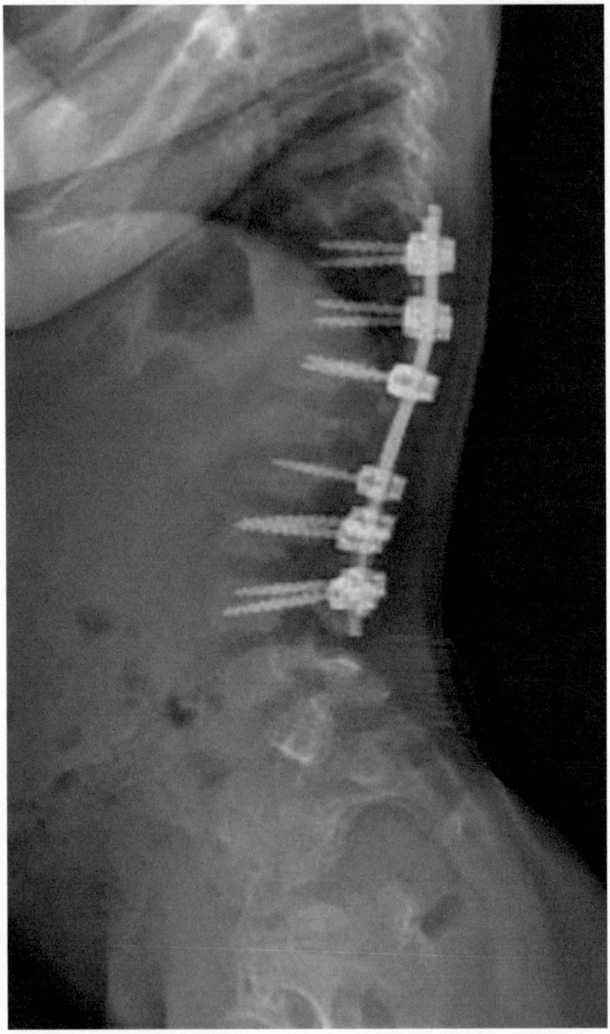

Fig. 15.3 The same child following a posterior spinal fusion T10–L4 with decompression at L3, L2, and L1

of neuromonitoring. The new baseline MEPs showed an anterior spinal cord dysfunction, as seen in the previous event, but the distal signals were improved, consistent with a resolving spinal cord injury. Suspecting only partial recovery from the previous injury to the spinal cord, the procedure was aborted again. He was brought back 8 weeks later. Again, he was kept in lateral position from the start. The blood pressure was supported with dopamine to a mean of 70–75 mmHg. Neuromonitoring showed that the trunk and lower extremity were now monitorable with double-train stimulation. He successfully underwent a posterior spinal fusion from T10–L4 with decompression at L3, L2, and L1 (Fig. 15.3). Seventeen months after his aborted hip dysplasia procedure, he was able to successfully undergo bilateral femoral and Dega osteotomies.

The above case illustrates the risk an MPS patient with a gibbus deformity faces when undergoing anesthesia, even for non-spinal procedures. The supine position can cause a stretch on the spinal cord over the kyphotic spine. Even the relatively brief time between induction and establishing neuromonitoring, in this case, was enough to compromise the patient's spinal cord. Unmonitorable signals were not consistent with his preoperative clinical presentation, and thus most likely represented a new dysfunction of his spinal cord. This deduction was proven when he returned to the OR and his signals were consistent with an improving spinal cord injury. Without neuromonitoring, this patient may have sustained a devastating injury.

References

1. Muenzer J. Early initiation of enzyme replacement therapy for the mucopolysaccharidoses. Mol Genet Metab. 2014;111:63–72.
2. Lorincz AE. The mucopolysaccharidoses: advances in understanding and treatment. Pediatr Ann. 1978;7:104–22.
3. D'Avanzo F, Rigon L, Zanetti A, Tomanin R. Mucopolysaccharidosis type II: one hundred years of research, diagnosis, and treatment. Int J Mol Sci. 2020;21:1258.
4. Çelik B, Tomatsu SC, Tomatsu S, Khan SA. Epidemiology of mucopolysaccharidoses update. Diagnostics (Basel). 2021;11:273.
5. Imundo L, Leduc CA, Guha S, Brown M, Perino G, Gushulak L, et al. A complete deficiency of hyaluronoglucosaminidase 1 (HYAL1) presenting as familial juvenile idiopathic arthritis. J Inherit Metab Dis. 2011;34:1013–22.
6. Natowicz MR, Short MP, Wang Y, Dickersin GR, Gebhardt MC, Rosenthal DI, et al. Clinical and biochemical manifestations of hyaluronidase deficiency. N Engl J Med. 1996;335:1029–33.
7. Zhou J, Lin J, Leung WT, Wang L. A basic understanding of mucopolysaccharidosis: incidence, clinical features, diagnosis, and management. Intractable Rare Dis Res. 2020;9:1–9.
8. NORD. Diseases. Mucopolysaccharidoses. 2017. https://rarediseases.org/rare-diseases/mucopolysaccharidoses/.
9. Muschol N, Giugliani R, Jones SA, Muenzer J, Smith NJC, Whitley CB, et al. MPS III Guideline Development Group. Sanfilippo syndrome: consensus guidelines for clinical care. Orphanet J Rare Dis. 2022;17:391.
10. Kampmann C, Abu-Tair T, Gökce S, Lampe C, Reinke J, Mengel E, et al. Heart and cardiovascular involvement in patients with mucopolysaccharidosis type IVA (Morquio-A syndrome). PLoS One. 2016;11:e0162612.
11. Caciotti A, Cellai L, Tonin R, Mei D, Procopio E, Di Rocco M, et al. Morquio B disease: from pathophysiology towards diagnosis. Mol Genet Metab. 2021;132:180–8.
12. NORD. Disorders. Maroteaux lamy syndrome. 2020. https://rarediseases.org/rare-diseases/maroteaux-lamy-syndrome
13. Montaño AM, Lock-Hock N, Steiner RD, Graham BH, Szlago M, Greenstein R, et al. Clinical course of sly syndrome (mucopolysaccharidosis type VII). J Med Genet. 2016;53:403–18.
14. Yeung AH, Cowan MJ, Horn B, Rosbe KW. Airway management in children with mucopolysaccharidoses. Arch Otolaryngol Head Neck Surg. 2009;135:73–9.
15. Gönüldaş B, Yılmaz T, Sivri HS, Güçer KŞ, Kılınç K, Genç GA, et al. Mucopolysaccharidosis: otolaryngologic findings, obstructive sleep apnea and accumulation of glucosaminoglycans in lymphatic tissue of the upper airway. Int J Pediatr Otorhinolaryngol. 2014;78:944–9.

16. Elwell Z, Mancuso D, Wolter NE, Propst EJ, Valdez T, Scheffler P. Post-tonsillectomy outcomes in children with mucopolysaccharidosis and obstructive sleep apnea. J Otolaryngol Head Neck Surg. 2023;52:87.
17. Braunlin EA, Harmatz PR, Scarpa M, Furlanetto B, Kampmann C, Loehr JP, et al. Cardiac disease in patients with mucopolysaccharidosis: presentation, diagnosis and management. J Inherit Metab Dis. 2011;34:1183–97.
18. Fesslová V, Corti P, Sersale G, Rovelli A, Russo P, Mannarino S, et al. The natural course and the impact of therapies of cardiac involvement in the mucopolysaccharidoses. Cardiol Young. 2009;19:170–8.
19. Spranger J. Bone dysplasias. 4th ed. Oxford: Oxford University Press; 2018.
20. McGrath O, Au L, Ashworth J. Management of corneal clouding in patients with mucopolysaccharidosis. J Clin Med. 2021;10:3263.
21. Taylor M, Khan S, Stapleton M, Wang J, Chen J, Wynn R, et al. Hematopoietic stem cell transplantation for mucopolysaccharidoses: past, present, and future. Biol Blood Marrow Transplant. 2019;25:e226–46.
22. Clarke LA, Giugliani R, Guffon N, Jones SA, Keenan HA, Munoz-Rojas MV, et al. Genotype-phenotype relationships in mucopolysaccharidosis type I (MPS I): insights from the International MPS I Registry. Clin Genet. 2019;96:281–9.
23. Clarke LA. Mucopolysaccharidosis Type I. 2002 Oct 31 [Updated 2024 Apr 11]. In: Adam MP, Feldman J, Mirzaa GM, et al., editors. GeneReviews® [Internet]. Seattle: University of Washington, Seattle; 1993–2024.
24. Fedele AO. Sanfilippo syndrome: causes, consequences, and treatments. Appl Clin Genet. 2015;8:269–81.
25. babysfirsttest.org. Baby's First Test. 2020. https://www.babysfirsttest.org/newborn-screening/about-babys-first-test
26. Langereis EJ, van Vlies N, Church HJ, Geskus RB, Hollak CE, Jones SA, et al. Biomarker responses correlate with antibody status in mucopolysaccharidosis type I patients on long-term enzyme replacement therapy. Mol Genet Metab. 2015;114:129–37.
27. Selvanathan A, Ellaway C, Wilson C, Owens P, Shaw PJ, Bhattacharya K. Effectiveness of early hematopoietic stem cell transplantation in preventing neurocognitive decline in mucopolysaccharidosis type II: a case series. JIMD Rep. 2018;41:81–9.
28. Sisinni L, Pineda M, Coll MJ, Gort L, Turon E, Torrent M, et al. Haematopoietic stem cell transplantation for mucopolysaccharidosis type VII: a case report. Pediatr Transplant. 2018;22:e13278.
29. Wood SR, Bigger BW. Delivering gene therapy for mucopolysaccharide diseases. Front Mol Biosci. 2022;9:965089.
30. Pruszczynski B, Mackenzie WG, Rogers K, White KK. Spinal cord injury after extremity surgery in children with thoracic kyphosis. Clin Orthop Relat Res. 2015;473:3315–20.
31. Society for Pediatric Anesthesia. Wake Up Safe. Spinal cord infarction or injury in patients with mucopolysaccharidosis. 2018. https://wakeupsafe.org/wp-content/uploads/2018/09/MPS_statement.pdf.
32. Tong CK, Chen JC, Cochrane DD. Spinal cord infarction remote from maximal compression in a patient with Morquio syndrome. J Neurosurg Pediatr. 2012;9:608–12.
33. Drummond JC, Krane EJ, Tomatsu S, Theroux MC, Lee RR. Paraplegia after epidural-general anesthesia in a Morquio patient with moderate thoracic spinal stenosis. Can J Anaesth. 2015;62:45–9.
34. Rosanò E, Tavoletti D, Luccarelli G, Cerutti E, Pecora L. Incidência de dispersão peridural após bloqueio do plexo lumbar com técnica de Chayen: estudo retrospectivo [Incidence of epidural spread after Chayen's approach to lumbar plexus block: a retrospective study]. Braz J Anesthesiol. 2020;70:202–8.

Osteogenesis Imperfecta

16

Manish Purohit and Divya Dixit

16.1 Introduction

Osteogenesis imperfecta (OI) is a rare skeletal dysplasia occurring in 1 in 15,000–20,000 births [1]. It is characterized by low bone density caused by qualitative and quantitative defects in the production of type I collagen [2].

Osteogenesis imperfecta was first reported in the 1840s, when William Vrolick described an infant with a poorly mineralized skeleton, large skull with frontal bossing, and thin bones making up the calvaria as well as the ribs. Vrolik was the first to postulate that this was an issue with primary ossification, not an acquired condition, leading to the term osteogenesis imperfecta. Later examination of the infant, who had survived only 3 days, revealed a diagnosis of OI type II [3].

In 1979, the Sillence classification defined a spectrum of four original types of OIs based on their presentation [4]. Today, with improved sequencing technologies, more than 20 types of OI are recognized with different affected genes. Osteogenesis imperfecta is now recognized as a spectrum of genetic disorders with tremendous variability in phenotypes, with both skeletal and extra skeletal manifestations challenging caregivers during the anesthetic and surgical management.

M. Purohit · D. Dixit (✉)
Department of Anesthesiology and Perioperative Medicine, Nemours Children's Hospital, Wilmington, Delaware, USA

Anesthesiology and Pediatrics, Sidney Kimmel Medical College, Thomas Jefferson University, Philadelphia, Pennsylvania, USA
e-mail: Manish.Purohit@nemours.org; Divya.Dixit@nemours.org

M. C. Theroux (ed.), *Anesthesia Care of Patients with Skeletal Dysplasia*,
https://doi.org/10.1007/978-3-031-92501-6_16

16.2 Genetics

Classically, OI is caused by defects in the production of type I collagen. Type I collagen forms a scaffolding for the formation and mineralization for the overlying bone. It is encoded by genes *COL1A1*, located on chromosome 17, and *COL1A2*, located on chromosome 7 [5]. Together, these genes code for the chains—two alpha 1 chains coded by *COL1A1* and one alpha 2 chain coded by *COL1A2*—that ultimately form the triple helix of type I collagen [6].

Approximately 90% of cases of OI arise from an autosomal dominant or sporadic mutation in the aforementioned alpha chains, which disrupts the helix [2, 7]. The most frequent cause of this classic form of OI is from glycine substitutions in the helix of the collagen. If these glycine substitutions are found in the alpha 1 chain, the phenotypes, including risk of lethality, are more severe than if the glycine substitutions are found in the alpha 2 helix [2].

The remaining 10% of OI cases arise from autosomal recessive or X-linked mutations [2]. With improvements in genetic testing over the last two decades, these recessive or X-linked forms have become evident and affect genes other than *COL1A1* and *COL1A2* [8]. The debate on the classification by genotype, including these variants not caused by abnormalities in the *COL1A1/1A2* genes, versus the classification by presenting phenotype remains unresolved.

16.3 Classification and Phenotype

Following a review of Vrolick's original patient description, Sillence and Dank in 1979 described four phenotypes of OI based on the classic autosomal dominant genotype, later found to be caused by defects in *COL1A1* and *COL1A2* genes (Table 16.1).

Over the last two decades, advances in sequencing have revealed phenotypes similar to the original four types described by Sillence but with completely unique genotypes and modes of inheritance. Furthermore, the phenotypes associated with the Sillence classification system may also vary thus creating further confusion [1]. Controversy continues with attempts to classify each genotype individually versus

Table 16.1 Sillence and Dank classification of osteogenesis imperfecta [9, 10]

Type	Features
I—mild	Blue sclerae, mild fragility, fractures after walking, no deformity
II—lethal	Blue sclerae, perinatal/in utero fractures, severe deformity, lethal
III—severe	Normal sclerae, very fragile, dentinogenesis imperfecta, scoliosis
IV—variable	Normal sclerae, variable fragility, and variable extra skeletal involvement

Table 16.2 Shapiro classification of osteogenesis imperfecta [12]

Type	Characteristics	Compatibility with life
Congenital A	In utero fractures of rib cage and long bones	Incompatible
Congenital B	In utero fractures but no rib cage deformity	Compatible
Tarda A	Fractures before walking	
Tarda B	Fractures after walking	

assigning each phenotype to the original classification regardless of genetic basis [10]. Further descriptors such as mild, moderate, and severe may also be helpful in offering a clearer picture of the extent of clinical disease, but this has not been standardized [11].

Other classification systems have attempted to delineate disease presentation based on prognosis. In 1906, Looser proposed separating patients into a congenital group and a tarda group. The congenital group suffered fractures occurring in utero or at birth; whereas, the tarda group's fractures occurred outside of the perinatal period. In 1985, Shapiro sought to refine this further by subdividing the Looser categories into A and B (Table 16.2) [12].

With these different systems, the nomenclature used by different authors to describe patients with OI can vary. However, the Sillence system, despite its perceived shortcomings, is the most utilized. As noted earlier, the phenotypic heterogeneity makes classification of the type of OIs confusing and difficult. Less severe forms tend to have fewer extra skeletal features while more severe forms tend to be more fragile with greater involvement of other organ systems (Table 16.3).

The above features are usually found in patients with more severe forms of OI and will warrant preoperative investigation and subsequent monitoring in the peri-anesthetic setting.

16.4 Anesthetic Concerns and Care

Patients with OI will likely undergo an anesthetic at some point in their lifetime. The variability of the phenotypic presentation of patients, coupled with the different surgeries, necessitates a wide range of anesthesia techniques, which may be concerning for anesthesiologists with less experience with this condition. To ensure patient safety for all procedures, certain considerations need to be addressed based upon the inherent risk of each procedure and the severity of the disease, as well as the resources available to the anesthesiologist. The issues and risks to be considered in patients with OI are listed in Table 16.4.

Table 16.3 Skeletal and extra skeletal features of osteogenesis imperfecta

Skeletal features	
Facial	Macrocephaly, triangular/elfin facies
	Frontal bossing, broad forehead
	Deep set eyes, beaked nose
Short stature	Abnormal development or healing of long bones
	Vertebral fracture/collapse
Spine [13]	Kyphoscoliosis
	Basilar impression/invagination
	Lumbosacral spondylolysis/ spondylolisthesis
Chest wall [14, 15]	Pectus excavatum/carinatum
Extra skeletal features	
Scleral hue	
Dentinogenesis Imperfecta [16]	
High pitched voice	
Joint hypermobility	
Pulmonary	Restrictive lung disease, tracheal narrowing and/or deviation
Cardiovascular [2]	Aortic root dilation, aortic dissection
	Valvular regurgitation, arrhythmias
Hearing loss	
Easy bruising	
Neurologic [17, 18]	Basilar invagination

Table 16.4 Peri-anesthetic risks and concerns in osteogenesis imperfecta

Difficult airway, upper airway obstruction, restrictive lung disease
Difficult intravenous access
Fracture from noninvasive blood pressure monitoring
Positioning related injuries
Neurologic injury from spinal anomalies
Coagulopathy
Hyperpyrexia
Postoperative pain control

16.5 Airway and Respiratory System

Patients with OI can present with difficult airway management with both difficult mask ventilation and tracheal intubation. While this is not common in patients with milder forms of the disease, the incidence is much higher in patients with severe types of OI [19] who are at risk for cervical spine instability from basilar invagination, and platybasia. The limited range of motion of the neck, further complicated by dentinogenesis imperfecta and chest wall deformity, may also contribute to the difficult airway.

Pulmonary compromise with decreased predicted vital capacity strongly correlated with thoracic scoliosis of more than 60° in the study by Widmann et al., who evaluated spine radiographs and pulmonary function tests in 15 adult patients with OI [13]. Pulmonary dysfunction was not correlated with kyphosis or chest wall deformity in this study [13].

Children with OI type III could develop vertebral collapse and scoliosis, which contribute to the restrictive lung disease. These skeletal features, as well as inactivity associated with wheelchair mobility, predispose to pneumonia. Restrictive lung pathophysiology with a mean FEV1/FVC ratio of 0.85 was found in a study of 42 patients with OI type III; the authors theorized that pulmonary impairment is a combined issue, partly associated to scoliosis and partly intrinsic to OI [20].

Sleep-disordered breathing and upper airway obstruction have been reported in children and neonates with OI [21, 22]. In the study by Leotard et al., 6.4% of the 188 patient charts studied had sleep-disordered breathing, and non-walkers had a higher apnea-hypopnea index than walkers, with two patients requiring continuous positive airway pressure therapy [21]. All of the neonates described in the three-case series by Johnson et al. had some degree of upper airway obstruction and restrictive lung disease, and two neonates required tracheostomy [22]. The upper airway obstruction may have an impact on emergence from anesthesia and the disposition of the patient after recovery from anesthesia.

Supraglottic airway placement may serve as a bridge for oxygenation and ventilation [23]; however, specialized equipment may be necessary to safely secure the airway. Video laryngoscopy may be used [24] and is commonly employed at our institution to intubate the trachea with minimal movement of the cervical spine, especially in patients with basilar invagination or other similar concerns [25]. Flexible bronchoscope is rarely necessary to secure the airway but should be readily available.

In our institution, we have noted occasional tracheal deviation in older patients with severe OI (Fig. 16.1). This anomaly is well documented in patients with mucopolysaccharidoses [26]. Smaller size of the endotracheal tube (ETT) and not using age-based ETT is appropriate and sufficient to ventilate and oxygenate the patients with OI, as they are much smaller in stature and weight when compared to similar in age, otherwise normal individuals.

16.6 Difficult Mask Ventilation

Though less likely in mild forms of OI, injuries or fractures to the bones of the face may occur during everyday activities, even yawning [27]. Mask ventilation, therefore, should be performed cautiously. When performing an inhalational induction, any significant pressure applied to the face to ensure mask seal may fracture the bridge of the nose, the alar prominences of the cheeks the maxilla, and the mandible, especially during the transition through the second stage of anesthesia. The anesthesia provider may consider obtaining an intravenous line thus avoiding inhalation induction.

If necessary, with difficult intravenous (IV) access or difficult airway, when the patient has reached an adequate depth of anesthesia, a laryngeal mask airway may

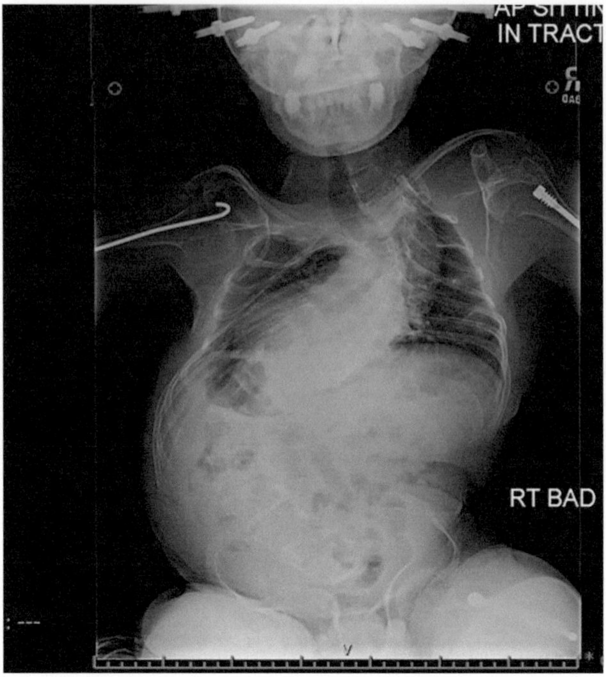

Fig. 16.1 A 20-year-old male with severe OI whose chest X-ray was difficult to interpret due to distortion of the chest. However, he was noted to have a deviated trachea on computed tomography/angiography of the chest. Note that the pins used for halo traction are visible

be placed to reduce the continuous pressure on the facial bones prior to intubation [23, 28]. As in most forms of skeletal dysplasia, the size of laryngeal mask airway may need to be smaller than expected for body weight.

16.7 Difficult Intravenous Access

The concern of the force applied by the cuff to the extremity in severe osteogenesis imperfecta with risk of iatrogenic fracture has been carried over to the use of a tourniquet in peripheral intravenous line access. If necessary, a tourniquet gently applied to the extremity or even manual circumferential pressure by hand can be used though rarely necessary.

Some patients may have an implanted port that can be used for induction of anesthesia. Younger patients may require sedative doses of nitrous oxide alone or in combination with sevoflurane to safely place an IV line. Difficult IV access is related to the severity of the type of OI; more than 50% of type III and IV OI patients are described as difficult. The overall incidence of difficult IV access was 4%, but the criteria for difficult IV access was not defined apart from case cancelation [19]. At our institution, difficult peripheral IV access is defined as need for ultrasound use or multiple placement attempts. Patients with severe forms of OI may present with shortened extremities or malformations from chronic fractures. Ultrasound guidance is recommended to secure IV access in all but mild cases of OI.

16.8 Use of Blood Pressure Cuff

The use of noninvasive BP cuff in patients with OI has typically been avoided due to the risk of iatrogenic fracture. Anecdotal evidence from families regarding the fragility of the patient and the fractures resulting from everyday activities [27] has driven the concern for iatrogenic fracture caused by the repeated cuff inflation from serial BP measurements. To avoid this potential complication, intra-arterial BP measurement has become the default choice for many anesthesiologists, with patients needing to be admitted to a critical care unit for hemodynamic monitoring, based on hospital policy.

Data from studies reviewing the incidence of fractures resulting from BP cuff are reassuring. Sullivan et al. reported no injuries in 81 procedures in patients with OI [29]. Similar results were noted in a retrospective study by Ross et al. in 2020; noninvasive BP measurement was used in 90% of cases of OI with no significant sequelae [30]. Some practitioners, however, rotated the BP cuff between extremities to decrease the frequency of inflation on a single extremity or placed padding underneath the BP cuff to further protect the extremity [30]. In addition, both studies reported no complications from arterial line placement at multiple sites depending upon the surgery.

In our experience, some patients with mild presentation of OI can tolerate noninvasive BP measurement thus avoiding the need for extra equipment and the risk associated with arterial cannulation. Those patients with more severe forms of OI may also tolerate noninvasive BP monitoring when the cuff is placed on a limb that has been surgically stabilized. Arterial line placement is deemed necessary based upon the severity of the medical condition, the potential for blood loss, and the duration of the surgery. Additionally, depending upon practitioner comfort, smaller gauge catheters may be placed intra-arterially for the procedure and removed at the end of the surgery or in the recovery area.

As the families have considerable experience in managing their child with OI, it is best to discuss preferences and departmental standards with them, the patient, and the surgeon to ensure patient comfort and safety.

16.9 Risk of Positioning Related Injury and Iatrogenic Fracture

Iatrogenic fracture is a risk in patients with severe OI, especially when positioning or moving the patient to and from the operating table, but it can be minimized [19, 31]. Preoperative discussion with the parents and the patient about the position that the patient typically prefers and the range of motion possible for extremities will guide optimal positioning and padding for appropriate support during the procedure. Using a draw sheet or other similar items along with sufficient support personnel may help with a smoother transfer. More importantly, clear communication amongst team members during each event can also mitigate the risk of an inadvertent movement that could injure the patient.

Historically, the prevention of iatrogenic fracture has directed the anesthetic management of patients with OI. As such, the avoidance of BP cuffs, succinylcholine and its induced fasciculation, and the avoidance of tourniquets have limited the management options. Iatrogenic fractures involving the face, skull, and extremities have also been reported related to body positioning [29].

16.10 Risk of Neurological Injury

Patients with OI are also at risk for neurological sequelae from craniocervical junction anomalies and spinal issues. Craniocervical anomalies consist of basilar invagination, basilar impression, and platybasia [17]. These conditions have been noted in as many as 37% of patients. While the anomalies may be present in every type of OI based on the Sillence classification, they occur more frequently in patients with severe types of OI. As the conditions may be asymptomatic, cervical spine X-rays are usually obtained before the age 6 of years, and further imaging with computed tomography or magnetic resonance imaging can be obtained as needed [32].

The anesthesiologist must take care in positioning patients by maintaining the cervical spine in a neutral to slightly extended position, based upon radiological findings and in discussion with the surgeon, to minimize the risk of spinal cord injury.

Spinal anomalies may also extend to the thoracic and lumbar spines in patients with OI. Scoliosis has been noted in all types of OI; though rarely before the age of 6 years [33]. The incidence varies greatly at 39–80% [13]. In milder types of OI, the thoracic spine is commonly affected. In more severe forms, however, scoliosis is less likely to involve only the thoracic spine and is often accompanied by more involved spinal anomalies [34]. The scoliosis is progressive and may warrant further imaging as the patient grows. Furthermore, as the scoliosis may progress rapidly, restrictive lung disease may lead to large alterations in vital capacity [13]; though it is unknown whether modern surgical correction techniques will prevent further decline [11].

For the anesthesiologist, timing of surgical correction varies from one institution to another, but in all cases of scoliosis correction, patients are monitored with transcranial motor evoked potentials, somatosensory evoked potentials, and electromyography. Use of neuromonitoring for non-spine surgeries in patients with OI who have a "spine at risk" is not defined and merits review with a neurosurgeon familiar with OI.[1]

[1] Please refer to Chap. 16: Anesthetic Management of Spine Fusion.

16.11 Hyperpyrexia Vs Malignant Hyperthermia

Patients with OI often have a core temperature that is slightly above normal. This is attributed to the hypermetabolism resulting from constant turnover of the cells from frequent fractures. An increase in core temperature accompanied by metabolic acidosis led to the diagnosis of malignant hyperthermia in one case report [35]. On follow-up testing of the patient, a caffeine halothane contracture test was negative. There is no evidence to suggest that patients with OI are at a higher risk of malignant hyperthermia; clinically, it is often observed that patients have a higher baseline temperature, which may further rise if aggressive temperature conservation measures are employed. Therefore, caution is exercised to not overheat patients during an anesthetic.

16.12 Coagulopathy

Osteogenesis imperfecta has been classically associated with increased intraoperative blood loss [36]. Multiple studies attempting to find a cause have not shown any treatable coagulopathy inherent to the disease [37–39]. Intraoperative blood loss was found to be associated with more severe forms of OI but likely is also related to the surgical technique [19]. With no clear cause noted, patients with severe forms of OI who are to undergo surgeries with anticipated high intraoperative blood loss should have blood available for possible transfusion. Antifibrinolytics such as tranexamic acid may be helpful to decrease surgical blood loss during more extensive surgical procedures.

16.13 Postoperative Pain Management[2]

Pain management in patients with OI may be difficult and is often compounded by the patient's chronic use of opioids and anxiety related to the upcoming surgery. Use of regional anesthesia is desirable. Therefore in our institution, peripheral nerve blocks are employed whenever possible. The technique may include a single shot administration of the local anesthetic or placement of a catheter for continuous local anesthetic infusion. We individualize the plan based on the patient's preoperative status, type of surgery, and anticipated need for postoperative pain control. Multimodal analgesic plans should include use of non-steroidal anti-inflammatory agents, acetaminophen, benzodiazepines, gabapentin, clonidine, and ketamine as indicated.

A limitation inherent to peripheral nerve blocks, especially the more advanced blocks, is the need for a dedicated pain management team, which may not always be available. Practitioners therefore have utilized neuraxial techniques in patients with OI. Neuraxial techniques have been well-established in pregnant patients with OI, and they also serve as an option for infants and children with OI [40–43].

[2] Please refer to Chap. 18: Pain Management Including Regional Anesthesia.

16.14 Other Concerns

Cardiac anomalies are documented in patients with OI and can manifest as arrhythmias, valvular anomalies, large vessel anomalies, and hypertension [44]. Valvular anomalies involving both the mitral and tricuspid valves were noted in children with *COL1A1* mutation as were arrhythmias including atrial fibrillation, bundle branch block, and premature ventricular complexes [45]. Furthermore, structural anomalies of the large vessels of the heart have also been noted. Though rare, increased diameter of the main pulmonary artery and aortic root dilatation have been associated with OI and can progress with age [45, 46]. As these anomalies may be found in all types of OI, electrocardiogram is usually completed as a screen during the initial investigation and monitored as needed. Consultation with cardiac anesthesiology and cardiology may be warranted to assess risk and optimize the patient.

16.15 Case Illustration

16.15.1 Case 1

A 13-year-old female weighing 28.5 kg with OI type III, scoliosis, and spondylolisthesis L5–S1 presented for posterior spine fusion with instrumentation, left rib resection, and cranial tongs placement (Fig. 16.2). She had history of more than 100 skeletal fractures, including eight fractures at birth. She had undergone approximately 25 surgeries at our institution, including bilateral femoral and tibial rod placement and rod placements in other long bones such as her humerus, radius, and ulna. She had reactive airway disease, which was well controlled. Cervical spine X-ray showed no abnormality. She had a history of postoperative nausea and vomiting along with a history of behavioral changes with morphine. She had been on treatment with pamidronate for her underlying disease.

Anesthesia was induced with intravenous ketamine, dexmedetomidine, and propofol boluses, following which tracheal intubation was performed using GlideScope (Verathon, Bothell, WA) with a cuffed 6.0-mm ETT. Two IV lines and an arterial line were placed under ultrasound guidance. Anesthesia was maintained with a total intravenous technique to facilitate neuromonitoring of the spinal cord function. Agents used were infusion of propofol with bolus dose of ketamine and methadone. Esmolol infusion was used to maintain mean arterial BP in the range of 60–70 mmHg for the surgical exposure and later tapered off. During the instrumentation part of the surgical procedure, small doses of vasopressor were necessary to maintain and support BP.

A tranexamic acid bolus (30 mg/kg) was followed by infusion of 10 mg/kg/h throughout the surgery, which lasted approximately 6 h. Blood loss was approximately 600 mL (30% of estimated blood volume) and urine output was 320 mL (1.8 mL/kg/h). Total fluids administered were 3.1 L, which included 250 mL of cell-saver blood and a unit of fresh frozen plasma. The trachea was extubated in a deep plane of anesthesia in the operating room at the end of procedure, when a small

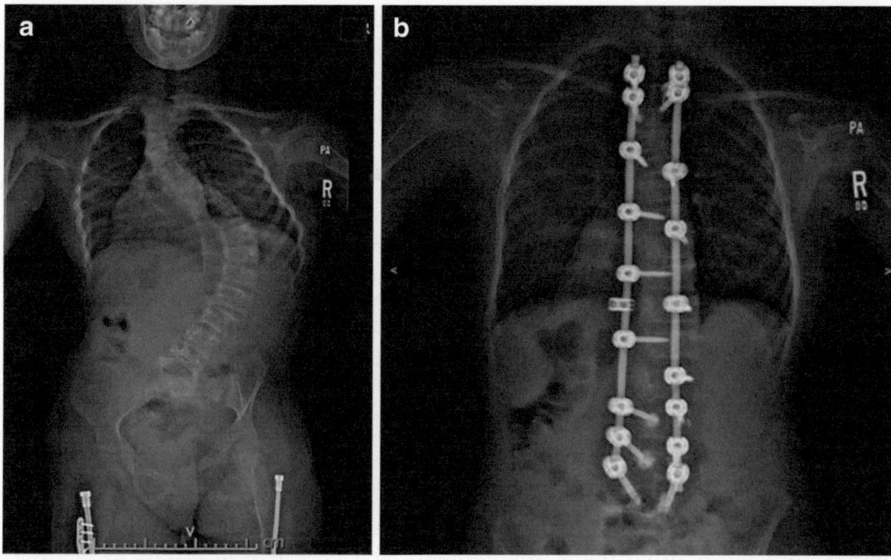

Fig. 16.2 X-ray image of the upright frontal view of the spine in the 13-year-old patient described in illustrative case 1, showing C-shaped curve with a thoracolumbar dextrocurvature and diffusely osteopenic bones (**a**). The thoracic kyphosis is corrected with posterior spine instrumentation in the X-ray image on the right (**b**)

dose of hydromorphone was administered as well. Care was taken to avoid any coughing or bucking during emergence.

The pain team, who had met the patient and the family in the preoperative room, managed the patient both in the recovery room and on the floor and continued managing until her discharge on postoperative day 3. Pain management regimen included IV acetaminophen 10 mg/kg scheduled every 6 h for 24 h then orally as needed; IV hydromorphone 0.3 mg administered as needed every 3 h until the patient was able to take oral oxycodone; and diazepam 2 mg orally administered every 6 h as needed for muscle spasms. In addition, clonidine patch 0.05 mg every 24 h was prescribed. On postoperative day 1, 0.5 mg/kg IV ketorolac was started every 6 h for 1–2 days. Per our institutional protocol, gabapentin 100 mg oral twice daily was administered as adjunct for pain management. The patient had overall satisfactory pain control in the postoperative period as assessed by patient, parents, and the pain team.

16.15.2 Case 2

A 22-year-old female weighing 42 kg with OI type III presented for left lower extremity osteotomy with realignment of intramedullary rod, removal of left femur hardware, and left femoral percutaneous skeletal fixation. She had associated preexisting conditions, including scoliosis and hydrocephalus. She had undergone multiple procedures in the past, including right femoral osteotomy with intramedullary rod, right open femoral fixation, and T5–L5 posterior spinal fusion. As is common for OI patients, she had sustained multiple bilateral lower extremity fractures with nonunion and undergone procedures to mitigate deformity and improve function. She had undergone a ventriculo-peritoneal shunt placement for hydrocephalus. She did not have any occiput–C1 collapse or gross malalignment but did have mild cervical kyphosis. She had ongoing pain related to her low back and neck. The almost daily headaches were associated with photophobia, phonophobia, and nausea, with presumptive diagnosis of migraine or occipital neuralgia. She had a left-sided drain placement for subdural hematoma related to a fall in the past. More recently, she had right-sided burr hole placement for intracranial pressure monitoring for progressive headaches. Due to a history of blurry vision with the use of scopolamine patch, the patient had requested to avoid its use for prevention of postoperative nausea and vomiting.

Anesthesia mask induction was initiated with inhalation agent using sevoflurane 8% and oxygen. Two peripheral IVs and an arterial line were placed. Tracheal intubation was accomplished after a small dose of non-depolarizing muscle relaxant using GlideScope, and a 6.5-mm cuffed ETT was placed. Total intravenous anesthesia was initiated with propofol, remifentanil, and lidocaine infusions to facilitate smooth emergence and prevent postoperative nausea and vomiting. A lumbar plexus catheter was placed on the left side after turning the patient carefully to the right lateral decubitus position (Fig. 16.3). After visualization of vertebra, transverse process, psoas muscle, quadratus lumborum muscle, and erector spinae muscles, 20 mL of 0.2% ropivacaine was administered with an 18 g insulated needle. Nerve stimulation during the regional anesthesia was avoided to prevent iatrogenic fractures. 10-mg methadone IV was administered in two divided doses to optimize pain management [47]. Anesthesia duration was about 4.5 h. Tranexamic acid 30 mg/kg bolus followed by infusion of 10 mg/kg/h was continued throughout surgery. Blood loss was <5 mL/kg, urine output was 495 mL (2.6 mL/kg/h), and the total

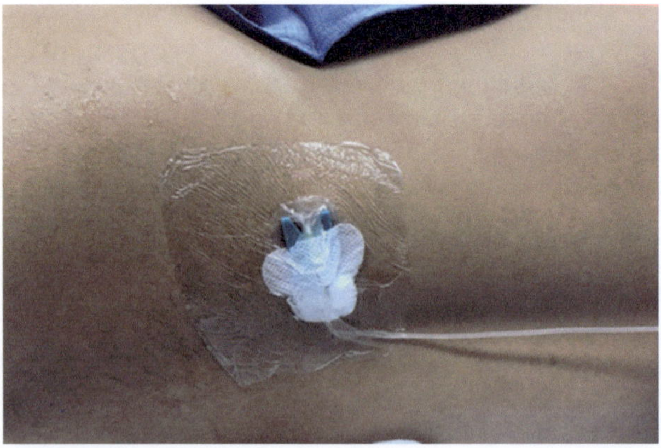

Fig. 16.3 Unilateral lumbar plexus catheter placed after induction of anesthesia prior to the start of surgery

crystalloid administered was 1100 mL. IV acetaminophen 15 mg/kg and ondansetron were administered toward the end of procedure. The trachea was extubated "deep" with the patient spontaneously breathing. A continuous infusion of local anesthetic with 0.2% ropivacaine was started via the lumbar plexus catheter in the recovery room and was discontinued on postoperative day 1 prior to discharge. Pain management also included IV morphine and ketorolac, oral diazepam, oxycodone, and acetaminophen.

The above cases illustrate two different patients undergoing extensive surgical procedures commonly performed for patients with OI. Note that the use of anesthetic agents such as ketamine and methadone are utilized due to patients' preoperative history of extensive skeletal fractures and chronic use of opioids. Use of regional techniques such as lumbar plexus block is desirable and requires skilled providers, especially considering the limitations placed on the anesthesia team by having to avoid safety measures such as nerve stimulation that they routinely practice during such procedures.

References

1. Bregou Bourgeois A, Aubry-Rozier B, Bonafé L, Laurent-Applegate L, Pioletti DP, Zambelli PY. Osteogenesis imperfecta: from diagnosis and multidisciplinary treatment to future perspectives. Swiss Med Wkly. 2016;146:w14322.
2. Forlino A, Marini JC. Osteogenesis imperfecta. Lancet. 2016;387:1657–71.
3. Baljet B. Aspects of the history of osteogenesis imperfecta (Vrolik's syndrome). Ann Anat. 2002;184:1–7.
4. Sillence DO, Rimoin DL, Danks DM. Clinical variability in osteogenesis imperfecta-variable expressivity or genetic heterogeneity. Birth Defects Orig Artic Ser. 1979;15:113–29.
5. Büttner C, Skupin A, Rieber EP. Transcriptional activation of the type I collagen genes COL1A1 and COL1A2 in fibroblasts by interleukin-4: analysis of the functional collagen promoter sequences. J Cell Physiol. 2004;198:248–58.
6. Scott JE, Haigh M. Proteoglycan-type I collagen fibril interactions in bone and non-calcifying connective tissues. Biosci Rep. 1985;5:71–81.
7. Barnes AM, Chang W, Morello R, Cabral WA, Weis M, Eyre DR, et al. Deficiency of cartilage-associated protein in recessive lethal osteogenesis imperfecta. N Engl J Med. 2006;355:2757–64.
8. Marini JC, Forlino A, Bächinger HP, Bishop NJ, Byers PH, Paepe A, et al. Osteogenesis imperfecta. Nat Rev Dis Primers. 2017;3:17052.
9. Sillence DO, Senn A, Danks DM. Genetic heterogeneity in osteogenesis imperfecta. J Med Genet. 1979;16:101–16.
10. Van Dijk FS, Sillence DO. Osteogenesis imperfecta: clinical diagnosis, nomenclature and severity assessment. Am J Med Genet A. 2014;164a:1470–81.
11. Franzone JM, Shah SA, Wallace MJ, Kruse RW. Osteogenesis imperfecta: a pediatric orthopedic perspective. Orthop Clin North Am. 2019;50:193–209.
12. Shapiro F. Consequences of an osteogenesis imperfecta diagnosis for survival and ambulation. J Pediatr Orthop. 1985;5:456–62.
13. Widmann RF, Bitan FD, Laplaza FJ, Burke SW, DiMaio MF, Schneider R. Spinal deformity, pulmonary compromise, and quality of life in osteogenesis imperfecta. Spine (Phila Pa 1976). 1999;24:1673–8.
14. Centeno NC, Justiniano VNO. Pectus excavatum in osteogenesis imperfecta type I treated with Nuss procedure after pamidronate therapy. J Pediatr Surg Case Reports. 2021;71:101915.

15. Orrick BA, Pierce AL, Snyder CL, Alon US. Successful brace treatment of pectus carinatum in osteogenesis imperfecta using the dynamic compression system. Eur J Pediatr Surg Rep. 2019;7:e117–20.
16. Xiao S, Yu C, Chou X, Yuan W, Wang Y, Bu L, et al. Dentinogenesis imperfecta 1 with or without progressive hearing loss is associated with distinct mutations in DSPP. Nat Genet. 2001;27:201–4.
17. Arponen H, Mäkitie O, Haukka J, Ranta H, Ekholm M, Mäyränpää MK, et al. Prevalence and natural course of craniocervical junction anomalies during growth in patients with osteogenesis imperfecta. J Bone Miner Res. 2012;27:1142–9.
18. Wadanamby S, El Garwany S, Connolly D, Arundel P, Bishop NJ, DeVile CJ, et al. Monitoring skull base abnormalities in children with osteogenesis imperfecta: review of current practice and a suggested clinical pathway. Bone. 2022;154:116235.
19. Rothschild L, Goeller JK, Voronov P, Barabanova A, Smith P. Anesthesia in children with osteogenesis imperfecta: retrospective chart review of 83 patients and 205 anesthetics over 7 years. Paediatr Anaesth. 2018;28:1050–8.
20. Keuning MC, Leeuwerke SJG, van Dijk PR, Harsevoort AGJ, Grotjohan HP, Franken AAM, et al. Prevalence of scoliosis and impaired pulmonary function in patients with type III osteogenesis imperfecta. Eur Spine J. 2022;31:2295–300.
21. Léotard A, Taytard J, Aouate M, Boule M, Forin V, Lallemant-Dudek P. Diagnosis, follow-up and management of sleep-disordered breathing in children with osteogenesis imperfecta. Ann Phys Rehabil Med. 2018;61:135–9.
22. Johnson K, Pineda M, Darrow D, Proud V, Derkay C. Neonatal upper airway obstruction in osteogenesis imperfecta: series of three cases and review of the literature. Ann Otol Rhinol Laryngol. 2008;117:27–31.
23. Sony S, Dalai M, John B, Sadhankar AD. I-gel as an interface before intubation to buy safe apnea time in a patient with anticipated difficult airway due to multiple facial fractures. Ain-Shams J Anesthesiol. 2023;15:95.
24. Khan FA, Shah S, Shah S, Mandoff V, Brooks GW. Use of GlideScope in airway management of a patient with osteogenesis imperfecta. J Anesth. 2012;26:310–1.
25. Singleton BN, Morris FK, Yet B, Buggy DJ, Perkins ZB. Effectiveness of intubation devices in patients with cervical spine immobilisation: a systematic review and network meta-analysis. Br J Anaesth. 2021;126:1055–66.
26. Averill LW, Kecskemethy HH, Theroux MC, Mackenzie WG, Pizarro C, Bober MB, et al. Tracheal narrowing in children and adults with mucopolysaccharidosis type IVA: evaluation with computed tomography angiography. Pediatr Radiol. 2021;51:1202–13.
27. Ram H, Shadab M, Vardaan A, Aga P. Fracture of mandible during yawning in a patient with osteogenesis imperfecta. BMJ Case Rep. 2014;2014:bcr2013203385.
28. Karabiyik L, Parpucu M, Kurtipek O. Total intravenous anaesthesia and the use of an intubating laryngeal mask in a patient with osteogenesis imperfecta. Acta Anaesthesiol Scand. 2002;46:618–9.
29. Sullivan BT, Margalit A, Garg VS, Njoku DB, Sponseller PD. Incidence of fractures from perioperative blood pressure cuff use, tourniquet use, and patient positioning in osteogenesis imperfecta. J Pediatr Orthop. 2019;39:e68–70.
30. Ross KE, Gibian JT, Crockett CJ, Martus JE. Perioperative considerations in osteogenesis imperfecta: a 20-year experience with the use of blood pressure cuffs, arterial lines, and tourniquets. Children (Basel). 2020;7(11):214.
31. Georgiadis AG, Pottinger TJ, Sigrist EJ, Truong WH. Iatrogenic intraoperative fracture in patients with osteogenesis imperfecta. J Pediatr Orthop. 2024;44:e945–50.
32. Khandanpour N, Connolly DJ, Raghavan A, Griffiths PD, Hoggard N. Craniospinal abnormalities and neurologic complications of osteogenesis imperfecta: imaging overview. Radiographics. 2012;32:2101–12.
33. Benson DR, Newman DC. The spine and surgical treatment in osteogenesis imperfecta. Clin Orthop Relat Res. 1981;159:147–53.

34. Anissipour AK, Hammerberg KW, Caudill A, Kostiuk T, Tarima S, Zhao HS, et al. Behavior of scoliosis during growth in children with osteogenesis imperfecta. J Bone Joint Surg Am. 2014;96:237–43.
35. Porsborg P, Astrup G, Bendixen D, Lund AM, Ording H. Osteogenesis imperfecta and malignant hyperthermia. Is there a relationship? Anaesthesia. 1996;51:863–5.
36. Persiani P, Pesce MV, Martini L, Ranaldi FM, D'Eufemia P, Zambrano A, et al. Intraoperative bleeding in patients with osteogenesis imperfecta type III treated by Fassier-Duval femoral rodding: analysis of risk factors. J Pediatr Orthop B. 2018;27:338–43.
37. Wong RS, Follis FM, Shively BK, Wernly JA. Osteogenesis imperfecta and cardiovascular diseases. Ann Thorac Surg. 1995;60:1439–43.
38. Langness U, Behnke H. Clinical picture and genetics of osteogenesis imperfecta. Dtsch Med Wochenschr. 1970;95:209–12; passim.
39. Gooijer K, Rondeel JMM, van Dijk FS, Harsevoort AGJ, Janus GJM, Franken AAM. Bleeding and bruising in osteogenesis imperfecta: international society on thrombosis and haemostasis bleeding assessment tool and haemostasis laboratory assessment in 22 individuals. Br J Haematol. 2019;187:509–17.
40. Onal O, Emin Zora M, Aslanlar E, Ozdemirkan A, Bengi CJ. Spinal anesthesia in an infant with osteogenesis imperfecta. Anesth Pain Med. 2018;8:e65917.
41. Gupta A, Kamal G, Gupta N, Aggarwal A. Combined spinal-epidural anesthesia with dexmedetomidine-based sedation for multiple corrective osteotomies in a child with osteogenesis imperfecta type III: a case report. A A Case Rep. 2017;9:60–3.
42. Murray S, Shamsuddin W, Russell R. Sequential combined spinal-epidural for caesarean delivery in osteogenesis imperfecta. Int J Obstet Anesth. 2010;19:127–8.
43. Yu R, Zhou J, Zhu Q. Pregnancy complicated by severe osteogenesis imperfecta poses a challenge for the anaesthetist: a case report. J Int Med Res. 2024;52(6):3000605241260551.
44. Verdonk SJE, Storoni S, Micha D, van den Aardweg JG, Versacci P, Celli L, et al. Is osteogenesis imperfecta associated with cardiovascular abnormalities? A systematic review of the literature. Calcif Tissue Int. 2024;114:210–21.
45. Zhao D, Liu Y, Liu J, Hu J, Zhang Q, Wang O, et al. Cardiovascular abnormalities and its correlation with genotypes of children with osteogenesis imperfecta. Front Endocrinol (Lausanne). 2022;13:1004946.
46. Balasubramanian M, Verschueren A, Kleevens S, Luyckx I, Perik M, Schirwani S, et al. Aortic aneurysm/dissection and osteogenesis imperfecta: four new families and review of the literature. Bone. 2019;121:191–5.
47. Kharasch ED, Brunt LM, Blood J, Komen H. Intraoperative methadone in next-day discharge outpatient surgery: a randomized, double-blinded, dose-finding pilot study. Anesthesiology. 2023;139:405–19.

Part III

Anesthesia for Spine Fusion and Neuromonitoring

17 Anesthetic Management of Spine Fusion

Mary C. Theroux, Kesavan Sadacharam,
and Lauren W. Averill

17.1 Introduction

Anesthesia for decompression and fusion of the spine may be broadly categorized into (1) anesthetic management of cervical spine decompression/fusion and (2) thoracolumbar spine decompression/fusion. There are many similarities between the two, but there are also some subtle differences.

The goals of anesthetic management are to (1) optimize intubation methods to avoid injury to the cervical spine cord, (2) optimize hemodynamic parameters to maximize evoked potentials, (3) use procoagulants to decrease blood loss, (4) optimize fluid management to minimize swelling and edema of oropharyngeal and laryngeal tissue, and (5) tailor the anesthetic management to achieve safe extubation in the operating room at the end of the procedure.

M. C. Theroux (✉) · K. Sadacharam
Department of Anesthesiology and Perioperative Medicine, Nemours Children's Hospital, Wilminton, Delaware, USA

Anesthesiology and Pediatrics, Sidney Kimmel Medical College, Thomas Jefferson University, Philadelphia, Pennsylvania, USA
e-mail: mtheroux@nemours.org; kesavan.sadacharam@nemours.org

L. W. Averill
Radiology and Pediatrics, Sidney Kimmel Medical College, Thomas Jefferson University, Philadelphia, Pennsylvania, USA

Department of Radiology, Nemours Children's Hospital, Wilmington, Delaware, USA
e-mail: Lauren.averill@nemours.org

M. C. Theroux (ed.), *Anesthesia Care of Patients with Skeletal Dysplasia*,
https://doi.org/10.1007/978-3-031-92501-6_17

Dysplasias with a high incidence of cervical spine abnormalities are achondroplasia, Larsen syndrome, hypochondroplasia, spondyloepiphyseal dysplasia, Kniest dysplasia, diastrophic dysplasia, atelosteogenesis type 3, metatropic dysplasia, pseudoachondroplasia, campomelic dysplasia, and some mucopolysaccharidoses (Morquio syndrome [MPS IVA], Maroteaux-Lamy syndrome [MPS V1], and Sanfilippo syndrome [MPS III]). In many dysplasias, cervical spine procedures may be necessary early in life (younger than 5 years). Documented abnormalities of the cervical spine include kyphotic curve and atlantoaxial instability. They are commonly present in Larsen syndrome, diastrophic dysplasia, pseudoachondroplasia, spondyloepiphyseal dysplasia, Kniest syndrome, metatropic dysplasia, and mucopolysaccharidosis—especially MPS IVA and Maroteaux-Lamy syndromes [1–4]. Other dysplasias are fraught with cervical spine abnormalities, but the aforementioned are some of the more common.

Thoracolumbar kyphoscoliosis is observed commonly in the same dysplasias in which cervical spine abnormalities are present. Anesthetic management of thoracolumbar spine fusion encompasses all the characteristics of cervical spine procedures, albeit with less emphasis and concern regarding cervical spine injury during airway management. Typically, thoracolumbar procedures are more extensive, often resulting in increased blood loss and prolonged surgical duration [5].

17.2 Airway

Many forms of skeletal dysplasia result in difficult airway management, which worsens with age [6]. Airway management in earlier life may be easier but often becomes progressively more difficult. Keeping the head and neck in a neutral position to minimize cervical spine movements is necessary and is a good practice to employ even when there is no overt cervical spine concerns. It is an excellent habit to develop where every patient with skeletal dysplasia is positioned on the operating table such that a shoulder roll aligns the clavicles with the external auditory meatus, as in Fig. 17.1.

A discussion with the surgical team will help discern the extent of cord compression that occurs with flexion and extension of the neck, as all such patients would have had their cervical spine imaged both in flexion and extension in addition to the neutral position. This invaluable study provides the necessary information related to occiput and cervical spine movement in the flexed and extended maneuvers [7]. At our institution, patients routinely have plain radiographs and MRI (when indicated) of the cervical spine performed in neutral position, flexion, and extension before their cervical spine procedures. These imaging studies are frequently performed with the child under anesthesia, in which case the radiologist will evaluate the initial scouting film in neutral head and neck position and determine the safety of flexion/extension maneuverability. A radiology technician then is instructed to flex and extend the patient's neck with an anesthesiologist in attendance.

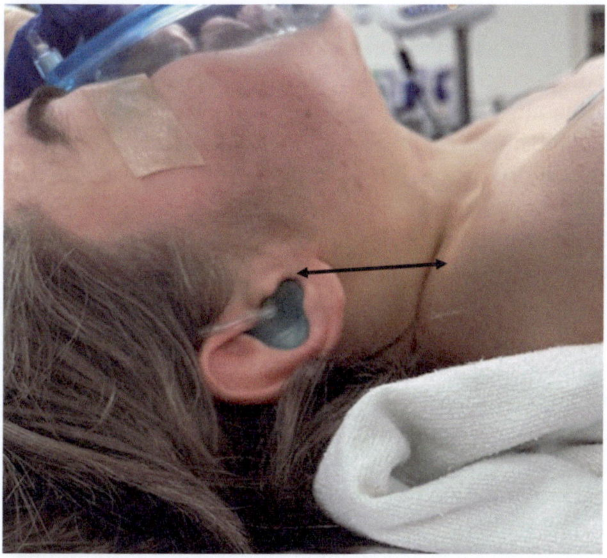

Fig. 17.1 A small shoulder roll is in place to align the external auditory meatus with the clavicle (arrow)

17.2.1 Airway Management

Utmost care is taken to prevent injuring the spinal cord with flexion or extension during tracheal intubation. Flexion is often observed as the more injurious motion when compared to extension, but keeping the head and neck in a neutral position is the best solution. Using the relatively new tools of intubation, video laryngoscope and flexible bronchoscope (which may also be combined for a hybrid technique), intubating the trachea without resulting in undue head and neck movements is not as problematic as it used to be. The patient's head and neck are maintained in the neutral position with a shoulder roll placed under the shoulder such that the external auditory meatus is aligned with the clavicle (Fig. 17.1). A provider, typically either an anesthesia provider or a provider from the surgical team, will maintain this neutral position during intubation. When using video laryngoscopes, the movement at the occiput to C1 and between C1 and C2 is minimized.

If enough concern exists because of the severity of the instability, pre- and post-intubation neuromonitoring can be added to enhance the safety of the intubation procedure. A typical such scenario would be a known patient with cervical spine instability with radiological evidence of spinal cord compression during flexion/extension. Preparations are made in conjunction with the surgeon to monitor the intubation procedure. Following a careful induction of anesthesia and placement of an intravenous line, the neurophysiologist will place enough electrodes to monitor both somatosensory and motor evoked potentials. A bite block is quickly placed to prevent tongue laceration during evoked potential stimulation, and baseline quickly is obtained. Following intubation, evoked potentials are obtained again and compared to baseline to ensure the spinal cord integrity.

Anesthetic drugs used to intubate the trachea require careful consideration, including avoiding muscle relaxants. In such patients, adjuvant drugs known to provide adequate intubating conditions (e.g., remifentanil) are preferred [8]. When examining adjuvant drugs to aid intubation (following induction using sevoflurane and/or propofol), Aouad et al. found remifentanil, lidocaine, and propofol most valuable [9]. Other studies have examined the use of drugs other than neuromuscular blockers to aid in intubation. Consistent results over multiple studies were observed with the combination of sevoflurane and/or propofol when remifentanil was added [10]. Increasing the dose of remifentanil from 1 to 3–4 μg/kg improved intubating conditions [11]. For more than 10 years, we have used intravenous remifentanil consistently in doses of 1–3 μg/kg following sevoflurane and propofol induction to intubate the trachea in our patients with skeletal dysplasia (SD) [4, 12]. The use of midazolam as a premedication is encouraged even though older patients often develop a dislike for the taste and choose to opt-out. Maintenance of anesthesia is accomplished with a combination of propofol and remifentanil [13–18], with small doses of adjuvants such as ketamine and/or dexmedetomidine. The goal is to extubate the trachea at the end of the procedure with the patient awake and able to obey commands.

For patients old enough to comprehend, a preoperative discussion of what to expect at the end of the procedure is highly recommended. For instance, informing patients that promptly taking a deep breath on command can expedite the removal of the endotracheal tube. Additionally, we encourage patients to demonstrate their ability to protrude their tongue on command. This action indicates sufficient muscle tone in the tongue, reducing the likelihood of tongue-based airway obstruction post-extubation. We routinely rehearse both actions (deep breathing and tongue protrusion) prior to anesthesia induction. This preparation significantly enhances the speed and efficiency with which patients can perform these tasks during the emergence phase.

Following spinal fusion, it is common for patients to exhibit varied degrees of facial swelling, including nasal mucosal engorgement. Simple interventions, such as applying a decongestant like Afrin (Bayer, Boca Raton, Florida) to the nostrils and preparing a high-flow oxygen delivery system (e.g., Vapotherm, Exeter, New Hampshire) prior to extubation, can significantly reduce post-extubation airway obstruction. The Vapotherm (Fig. 17.2) system fills the patient's respiratory dead space with fresh gas, facilitating an easier and less obstructed breathing pattern, particularly immediately after extubation. While these measures might appear theoretical, in practical application, they enhance safety, especially in high-risk cases.

17.2.2 Intubation

The need for maintaining a neutral position of the cervical spine especially in those patients with cervical spine instability has been emphasized already. Different

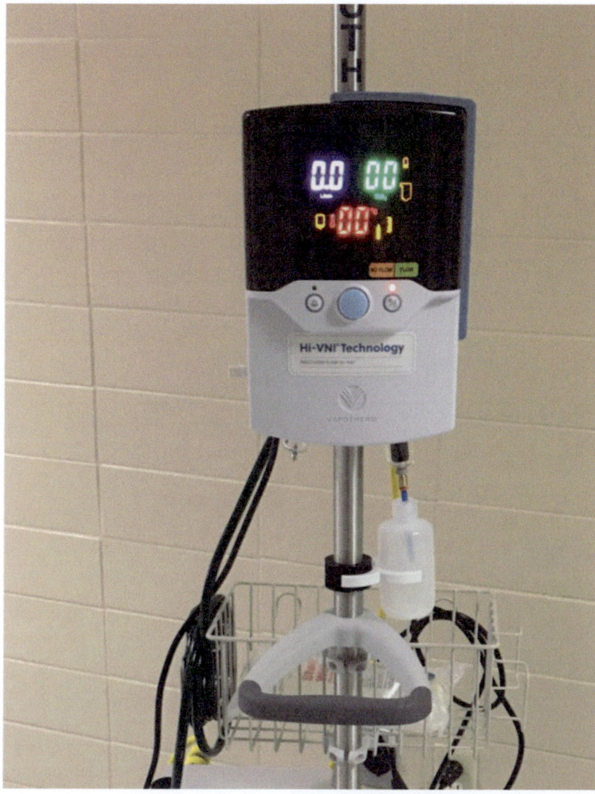

Fig. 17.2 Vapotherm is a non-invasive respiratory assistance device, effective in decreasing work-of-breathing, improving ventilation, and aiding in the recovery phase of patients with respiratory compromise

cervical spine levels were examined for movements during intubation employing various indirect techniques, primarily video laryngoscope [19–25]. A study examining cervical spine motion during conventional intubation technique using a Macintosh blade, one of the most commonly used tools across anesthesia departments [26], used continuous lateral fluoroscopy to assess movements at different cervical spine levels. They demonstrated that the visualization of the larynx using the Macintosh blade resulted in superior rotation of the occiput and C1 in the sagittal plane and mild inferior rotation of C3–C5, with C2 maintaining an almost neutral position. In addition, the report added that the most segmental movement occurred at atlanto-occipital and atlantoaxial levels.

Difficulty with airway management varies between different dysplasia and within the same dysplasia between patients as well. As patients become older, the airway often becomes more challenging to manage. Appearance and the anatomical features of head and neck change and there is a commonly observed progression of

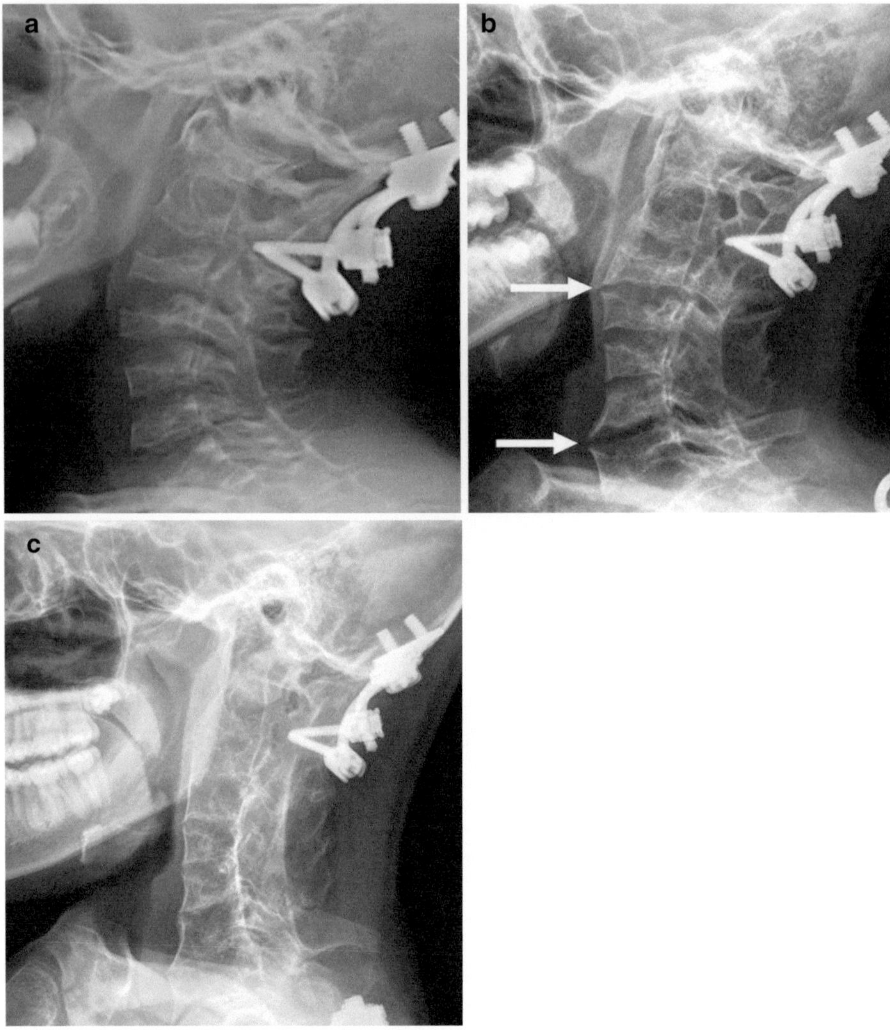

Fig. 17.3 (**a**) Lateral view of the cervical spine imaging in a girl with metatropic dysplasia at age 4, showing platyspondyly with occipital to C2 posterior spinal fusion. (**b**) Lateral view performed at 9 years old shows mature bony fusion from the occiput to C3 and an additional fusion of posterior elements from C4–C6. There is anterior beaking of vertebrae between these fused segments due to motion (arrows). (**c**) Lateral view at 16 years old shows complete bony fusion of the occiput and entire cervical spine with normal cervical lordosis. Clinically, this patient had no motion of her head and neck, rendering intubation of trachea difficult

bony abnormalities that lead to ever increasing difficulty with the airway management (Fig. 17.3). Many patients develop restricted mouth opening, which tends to be progressive. All these considerations need to be taken into account when planning airway management.

17.2.3 Anesthetic Agents

The choice of anesthetic agents should fit the goals detailed before, and the plan to extubate the patient's trachea at the end of the surgical procedure should be a significant consideration. An awake extubation is preferred to enhance safety by ensuring the least amount of tongue-based obstruction. The most compelling argument for extubating patients in the operating room is that, given the rate of airway difficulties in patients with SD, management of the airway, including extubation, is best performed in the operating room setting with the skilled anesthesiologist. Planning for a timely extubation, shorter acting anesthetic agents are preferred over longer-acting agents. Another major determinant of the anesthetic agents is the need for neuromonitoring. Continuous infusions of propofol and remifentanil may be considered the mainstay of the anesthetic agents. Additional agents are used (albeit sparingly) to ensure analgesia is on board when the patient emerges. Preferably, these agents are given as a bolus earlier during the anesthetic to ensure emergence with good airway control. Ketamine is most commonly used as a bolus and has the additional advantage of being a good bronchodilator as many patients with skeletal dysplasia have reactive airway disease or have a component of obstructive airway disease in addition to their restrictive airway disease [27, 28]. Narcotics should be minimized as much as possible in favor of non-steroidal analgesics, such as intravenous acetaminophen and or ketorolac. Other beneficial medications, such as benzodiazepines (namely diazepam), are best reserved for the recovery period when the residual anesthetics have mostly worn off and the patient has regained adequate tone of the airway and oropharyngeal muscles. Medications often discussed in the literature (i.e., methadone or other long-acting agents) have a limited role in the care of these complex patients with skeletal dysplasia. Another caveat mentioned in the literature is the hyperalgesia remifentanil may induce when used as a sole agent without other longer-acting analgesics. This essentially is a theoretical concern and clinically not observed unless remifentanil is used as a sole narcotic agent without using a regional block or another analgesic agent (e.g., concomitant administration of ketamine) on board [29].

There is a greater incidence of changes in evoked potentials indicative of an evolving neurological injury during surgery and anesthesia for patients with skeletal dysplasia [2, 4, 30–32]. In addition, changes in evoked potential monitoring have been reported in relation to the position of the patient. In this report of three cases, one patient improved with flexion of the head and neck, and the other two patients had to be repositioned multiple times, including one patient who had to be positioned awake [30]. The overall anesthetic plan should also consider the need for raising mean arterial pressures to optimize the monitoring of evoked potentials should there be a decrease or change in evoked potentials indicative of evolving or impending neurological injury.

Ketamine is an excellent bronchodilator and a supporter of blood pressure and is a well-known agent in decreasing the use of perioperative opioids [33]. It possesses anti-inflammatory properties as well [28], though the choice of ketamine as an adjuvant to other anesthetic agents is not directly related to its anti-inflammatory properties. Another useful adjuvant is dexmedetomidine [34], a selective alpha-2 agonist known for its lack of neurotoxicity and respiratory depression. This combination of drugs allows the patients to emerge with some analgesia on board but still awake for extubation. Dexmedetomidine has been examined for its possible effect on motor evoked potentials during spine fusion. It may attenuate the motor evoked potential by diminishing the amplitude when given in combination with propofol, administered at anesthetic concentrations. Dosing of propofol, when used in conjunction with dexmedetomidine, tends to be such that the depressive effects on neuromonitoring are not observed or not significant [35].

17.2.4 Procoagulants

Use of procoagulants are discussed and reported in multiple studies. The current procoagulant of choice is tranexamic acid, which has been studied extensively and has shown consistent results in reducing blood loss and transfusion needs, albeit at variable dosing regimens [36–38]. Other helpful adjuvants include preemptively treating reactive airway disease using methylprednisolone (2 mg/kg up to a maximum of 80 mg); for patients who have a history of reactive airway disease and/or have a reversible component of FEV1 in their spirometry testing, to prevent bronchospasm [39]. This applies to patients recovering from an upper respiratory infection as well, as the most common complication of an upper respiratory infection when superimposed on an anesthetic is bronchospasm.

17.2.5 Important to Remember

The airway should be prioritized when choosing anesthetic agents to allow safe and awake emergence at the end of the surgical procedure. Deep extubation may not be ideal and not recommended for a majority of patients with SD, especially when airway difficulties are encountered during intubation. Furthermore, the prevalence of obstructive sleep apnea in many forms of SD also plays a role in increased complications in the postoperative period. Tongue-based obstruction is also more common and should be expected in nearly every patient.

Some of the patients with SD may have a tracheostomy in place, and the need for securing the airway may involve suturing the endotracheal tube inserted via the tracheostomy. We have found the endotracheal tube to be most secure when sutured in place, especially as these patients are in prone position for many hours and taped-in-place endotracheal tubes often can become loose and dislodged (Fig. 17.4). The distance from the stoma of the tracheostomy to the mid trachea is very short in most of these patients, contributing to easier outward prolapsing of the balloon of the endotracheal tube and potential loss of airway. While as anesthesiologists, suturing endotracheal tubes are not a routine or even a common practice, it is remarkably easy to do so.

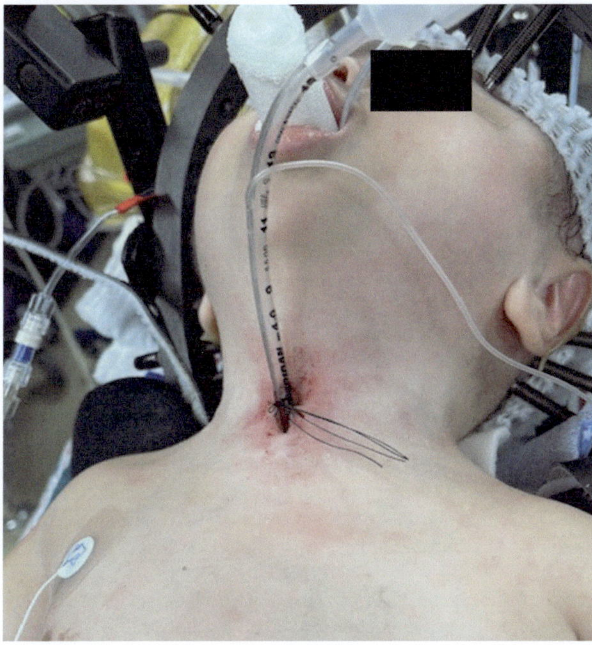

Fig. 17.4 Figure shows a patient with skeletal dysplasia who has an endotracheal tube inserted via a tracheostomy stoma. Tip of the endotracheal tube was positioned using a flexible bronchoscope and sutured in place in preparation for prone positioning, prior to a posterior spine fusion

17.3 Case Illustration

These medical imaging studies illustrate a typical patient with metatropic dysplasia and highlight the progressive cervical spine abnormalities (Fig. 17.3). This patient became progressively difficult to intubate due to the progression of her dysplasia. Obvious is the occipito-cervical fusion, but note the spontaneous fusion of the cervical spine below with loss of definition of the vertebral structures of the remaining cervical spine.

References

1. Thorne JA, Javadpour M, Hughes DG, Wraith E, Cowie RA. Craniovertebral abnormalities in type VI mucopolysaccharidosis (Maroteaux-Lamy syndrome). Neurosurgery. 2001;48:849–52; discussion 852–3.
2. Solanki GA, Martin KW, Theroux MC, Lampe C, White KK, Shediac R, et al. Spinal involvement in mucopolysaccharidosis IVA (Morquio-Brailsford or Morquio A syndrome): presentation, diagnosis and management. J Inherit Metab Dis. 2013;36:339–55.
3. Baratela WA, Bober MB, Thacker MM, Belthur MV, Oto M, Rogers KJ, et al. Cervicothoracic myelopathy in children with morquio syndrome a: a report of 4 cases. J Pediatr Orthop. 2014;34:223–8.

4. Theroux MC, Lopez M, Olszewsky PJ, DiCindio S, Arai L, Ditro C, et al. Metatropic dysplasia-a skeletal dysplasia with challenging airway and other anesthetic concerns. Paediatr Anaesth. 2017;27:596–603.
5. Carabini LM, Zeeni C, Moreland NC, Gould RW, Avram MJ, Hemmer LB, et al. Development and validation of a generalizable model for predicting major transfusion during spine fusion surgery. J Neurosurg Anesthesiol. 2014;26:205–15.
6. Shrader MW, DiCindio S, Kenny KG, Franco AJ, Zhang R, Theroux MC, et al. Transcranial electric motor evoked potential monitoring during scoliosis surgery in children with cerebral palsy and active seizure disorder: is it feasible and safe? Spine Deform. 2023;11:1461–6.
7. Mackenzie WG, Dhawale AA, Demczko MM, Ditro C, Rogers KJ, Bober MB, et al. Flexion-extension cervical spine MRI in children with skeletal dysplasia: is it safe and effective? J Pediatr Orthop. 2013;33:91–8.
8. Aouad MT, Yazbeck-Karam VG, Mallat CE, Esso JJ, Siddik-Sayyid SM, Kaddoum RN. The effect of adjuvant drugs on the quality of tracheal intubation without muscle relaxants in children: a systematic review of randomized trials. Paediatr Anaesth. 2012;22:616–26.
9. Aouad MT, Sayyid SS, Zalaket MI, Baraka AS. Intravenous lidocaine as adjuvant to sevoflurane anesthesia for endotracheal intubation in children. Anesth Analg. 2003;96:1325–7.
10. Klemola UM, Hiller A. Tracheal intubation after induction of anesthesia in children with propofol—remifentanil or propofol-rocuronium. Can J Anaesth. 2000;47:854–9.
11. Weber F, Füssel U, Gruber M, Hobbhahn J. The use of remifentanil for intubation in paediatric patients during sevoflurane anaesthesia guided by bispectral index (BIS) monitoring. Anaesthesia. 2003;58:749–55.
12. Theroux MC, DiCindio S, Averill LW, Pizarro C, Oommen A, Bober MB, et al. Tracheal narrowing and its impact on anesthesia care in patients with morquio A (mucopolysaccharidosis type IVA): an observational study. Anesth Analg. 2023;137:1075–83.
13. McPherson RW, Mahla M, Johnson R, Traystman RJ. Effects of enflurane, isoflurane, and nitrous oxide on somatosensory evoked potentials during fentanyl anesthesia. Anesthesiology. 1985;62:626–33.
14. Ravussin P, de Tribolet N, Wilder-Smith OH. Total intravenous anesthesia is best for neurological surgery. J Neurosurg Anesthesiol. 1994;6:285–9.
15. Langeron O, Vivien B, Paqueron X, Saillant G, Riou B, Coriat P, et al. Effects of propofol, propofol-nitrous oxide and midazolam on cortical somatosensory evoked potentials during sufentanil anaesthesia for major spinal surgery. Br J Anaesth. 1999;82:340–5.
16. Pelosi L, Stevenson M, Hobbs GJ, Jardine A, Webb JK. Intraoperative motor evoked potentials to transcranial electrical stimulation during two anaesthetic regimens. Clin Neurophysiol. 2001;112:1076–87.
17. Laureau E, Marciniak B, Hébrard A, Herbaux B, Guieu JD. Comparative study of propofol and midazolam effects on somatosensory evoked potentials during surgical treatment of scoliosis. Neurosurgery. 1999;45:69–74; discussion 75.
18. Scheufler KM, Zentner J. Total intravenous anesthesia for intraoperative monitoring of the motor pathways: an integral view combining clinical and experimental data. J Neurosurg. 2002;96:571–9.
19. Mentzelopoulos SD, Tzoufi MJ, Papageorgiou EP. The disposition of the cervical spine and deformation of available cord space with conventional- and balloon laryngoscopy-guided laryngeal intubation: a comparative study. Anesth Analg. 2001;92:1331–6.
20. Waltl B, Melischek M, Schuschnig C, Kabon B, Erlacher W, Nasel C, et al. Tracheal intubation and cervical spine excursion: direct laryngoscopy vs. intubating laryngeal mask. Anaesthesia. 2001;56:221–6.
21. LeGrand SA, Hindman BJ, Dexter F, Weeks JB, Todd MM. Craniocervical motion during direct laryngoscopy and orotracheal intubation with the Macintosh and Miller blades: an in vivo cinefluoroscopic study. Anesthesiology. 2007;107:884–91.
22. Hindman BJ, Santoni BG, Puttlitz CM, From RP, Todd MM. Intubation biomechanics: laryngoscope force and cervical spine motion during intubation with macintosh and airtraq laryngoscopes. Anesthesiology. 2014;121:260–71.

23. Suppan L, Tramèr MR, Niquille M, Grosgurin O, Marti C. Alternative intubation techniques vs macintosh laryngoscopy in patients with cervical spine immobilization: systematic review and meta-analysis of randomized controlled trials. Br J Anaesth. 2016;116:27–36.
24. Singleton BN, Morris FK, Yet B, Buggy DJ, Perkins ZB. Effectiveness of intubation devices in patients with cervical spine immobilisation: a systematic review and network meta-analysis. Br J Anaesth. 2021;126:1055–66.
25. Correa JBB, Felice VB, Sbruzzi G, Friedman G. Cervical spine movements during laryngoscopy and orotracheal intubation: a systematic review and meta-analysis. Emerg Med J. 2023;40:300–7.
26. Sawin PD, Todd MM, Traynelis VC, Farrell SB, Nader A, Sato Y, et al. Cervical spine motion with direct laryngoscopy and orotracheal intubation. An in vivo cinefluoroscopic study of subjects without cervical abnormality. Anesthesiology. 1996;85:26–36.
27. Loftus RW, Yeager MP, Clark JA, Brown JR, Abdu WA, Sengupta DK, et al. Intraoperative ketamine reduces perioperative opiate consumption in opiate-dependent patients with chronic back pain undergoing back surgery. Anesthesiology. 2010;113:639–46.
28. Loix S, De Kock M, Henin P. The anti-inflammatory effects of ketamine: state of the art. Acta Anaesthesiol Belg. 2011;62:47–58.
29. Blacker SN, Vincent A, Burbridge M, Bustillo M, Hazard SW, Heller BJ, et al. Society for neuroscience in anesthesiology and critical care. Perioperative care of patients undergoing major complex spinal instrumentation surgery: clinical practice guidelines from the Society for neuroscience in anesthesiology and critical care. J Neurosurg Anesthesiol. 2022;34:257–76.
30. Ofiram E, Lonstein JE, Skinner S, Perra JH. "The disappearing evoked potentials": a special problem of positioning patients with skeletal dysplasia: case report. Spine (Phila Pa 1976). 2006;31:E464–70.
31. Borlot F, Arantes PR, Capel Cardoso AC, Kim CA. Remote spinal cord injury in mucopolysaccharidosis type IVA after cervical decompression. Neurology. 2014;82:1382–3.
32. Sahin MC, Ipek EM, Zinnuroglu M, Borcek AO. How did intraoperative neuromonitorization prevent tetraplegia? Childs Nerv Syst. 2023;39:1323–8.
33. Schwenk ES, Viscusi ER, Buvanendran A, Hurley RW, Wasan AD, Narouze S, et al. Consensus guidelines on the use of intravenous ketamine infusions for acute pain management from the American Society of Regional Anesthesia and Pain Medicine, the American Academy of Pain Medicine, and the American Society of Anesthesiologists. Reg Anesth Pain Med. 2018;43:456–66.
34. Bala E, Sessler DI, Nair DR, McLain R, Dalton JE, Farag E. Motor and somatosensory evoked potentials are well maintained in patients given dexmedetomidine during spine surgery. Anesthesiology. 2008;109:417–25.
35. Tobias JD, Goble TJ, Bates G, Anderson JT, Hoernschemeyer DG. Effects of dexmedetomidine on intraoperative motor and somatosensory evoked potential monitoring during spinal surgery in adolescents. Paediatr Anaesth. 2008;18:1082–8.
36. Goobie S. The case for the use of tranexamic acid. Paediatr Anaesth. 2013;23:281–4.
37. Goobie SM, Meier PM, Sethna NF, Soriano SG, Zurakowski D, Samant S, et al. Population pharmacokinetics of tranexamic acid in paediatric patients undergoing craniosynostosis surgery. Clin Pharmacokinet. 2013;52:267–76.
38. Rahmani R, Singleton A, Fulton Z, Pederson JM, Andreshak T. Tranexamic acid dosing strategies and blood loss reduction in multilevel spine surgery: a systematic review and network meta-analysis: tranexamic acid for multilevel spine surgery. N Am Spine Soc J. 2021;8:100086.
39. Dewachter P, Mouton-Faivre C, Emala CW, Beloucif S. Case scenario: bronchospasm during anesthetic induction. Anesthesiology. 2011;114:1200–10.

Skeletal Dysplasia and Intraoperative Neuromonitoring

18

Sabina Dicindio and Anthony DiNardo

18.1 Introduction

The osteochondrodysplasias, also known as skeletal dysplasias (SDs), are characterized by cartilage and bone growth abnormalities. These abnormalities are genetic in origin and can result in abnormal shape, length, or density of the bones. The individual dysplasias are rare but represent 5% of children born with a birth defect, with an incidence of about 1 in 5000 live births [1]. A recent update to the classification of skeletal dysplasias places the number of clinical entities at 461 different diseases [2].

Embryologic bone development is from endochondral and membranous ossification, which is mesenchymal in origin, and forms the appendicular and axial skeleton, respectively. Development of the corticospinal motor tracts, somatosensory tracts, and peripheral nerves is primarily formed from ectodermal embryonic origin [3]. While the dysplasias are characterized by continual progression of disease, the nervous system is typically spared from ongoing degeneration of the neural tracts and pathways. Therefore, patients with SD typically have reliable evoked potentials (EPs) and other electrophysiological signals, allowing a baseline status to be established.

Patients with SD have a higher-than-normal risk of neural compromise due to a lifelong progression of deformity; the deformities and degree of severity are unique

S. Dicindio (✉)
Anesthesiology and Pediatrics, Sidney Kimmel Medical College,
Thomas Jefferson University, Philadelphia, Pennsylvania, USA

Department of Anesthesiology and Perioperative Care, Nemours Children's Hospital,
Wilmington, Delaware, USA
e-mail: sabina.dicindio@nemours.org

A. DiNardo
Department of Specialty Care, Nemours Children's Hospital, Wilmington, Delaware, USA
e-mail: Tony.DiNardo@nemours.org

M. C. Theroux (ed.), *Anesthesia Care of Patients with Skeletal Dysplasia*,
https://doi.org/10.1007/978-3-031-92501-6_18

to each patient. Most patients with SD have associated spine disorders, which present early in life and are often progressive [4]. The spine disorders include, but are not limited to, atlanto-axial instability, central stenosis, kyphosis, scoliosis, kyphoscoliosis, vertebral instability, and hyperlordosis. Due to the spine disorders, these patients are at increased risk for peripheral and central compression of the neural elements, which can lead to progressive myelopathy [4]. The patients may remain asymptomatic or present with clinical signs, such as central apnea, poor balance, unsteady gait, muscle weakness, or bowel and bladder dysfunction. Currently, therapeutic interventions are mostly supportive; however, many issues require surgical intervention, which is reported at 84.1% and is mostly for pain and neurologic symptoms [4, 5].

Due to pre-existing deformities and potential neural compromise, patients with SD often are considered to have a "spine-at-risk" [6]. For the same reason, surgery and anesthesia may result in new or worse postoperative neurologic deficit [6, 7]. The "spine-at-risk" patient is at greater risk for neural compromise with spinal corrective surgeries as well as non-spinal procedures. Mechanisms of neural compromise include direct stretching or compression of the cord, direct trauma from instrumentation, or interference with cord blood flow [8, 9]. Improper positioning of the head, neck, or shoulders, as well as hypotension, vasculopathy, and kyphosis are all factors that may contribute to spinal cord or neural injury [10–12]. Though spinal cord injury can occur at any time, head positioning is a particularly high-risk period [13]. There are several case reports involving spinal cord compromise associated with positioning during spine and non-spine surgeries [14–16]. Loss of evoked potentials with general anesthesia alone or in combination with neuraxial anesthesia also has been described [16, 17]. The area of greatest risk is thought to be at the T2–T10 thoracic levels due to the rigidity of the skeletal deformity. In addition, the anterior spinal artery (ASA) supplies blood to the anterior two-thirds of the spinal cord and receives collateral flow from radicular arteries. The ASA not only narrows in diameter through the thoracic spinal cord, but the number of radicular arteries is fewest at this level. For these reasons, this area of the thoracic spinal cord is known as a watershed area and is sensitive to cord ischemia [18–20]. Patients with increased lumbar lordosis or a horizontal sacrum also have a higher intraoperative complication rate attributed to positioning. This is attributed to decreased pelvic motion and malalignment of the muscle vectors [10, 21].

Disorders of the spine increase morbidity and mortality associated with any surgical procedure or anesthesia the patients undergo [4, 6, 7]. A major complication of surgical correction of scoliosis, kyphosis, or an unstable spine is potential injury to the spinal cord resulting in neurological deficit. While neurologic compromise secondary to spinal deformity correction is rare—0.14–0.79% in idiopathic spinal correction—patients with congenital stenosis or pre-existing neurologic abnormalities are at an increased risk for iatrogenic injury [9]. The pediatric SD population has a greater incidence of intraoperative monitoring changes when compared to those with infantile or juvenile scoliosis [4, 22]. In addition, the literature reports that patients with SD have a 12–23% incidence of new postoperative deficit after spinal corrective surgery that did not resolve at follow-up [23].

Intraoperative neuromonitoring (IONM) is utilized to monitor the integrity of the spinal cord while identifying and potentially preventing iatrogenic injury to the nervous system during spinal, neurological, cranial, and vascular procedures. Intraoperative neuromonitoring is a multimodal technique that includes somatosensory evoked potentials (SSEP), motor evoked potentials (MEP), electroencephalography (EEG), electromyography (EMG), and brainstem auditory evoked response (BAER). The Scoliosis Research Society states that intraoperative spinal cord monitoring facilitates detection of impending spinal cord injury, allowing early intervention that is likely to preserve spinal cord function [9, 10, 24–26]. Best practice guidelines published in 2020 recommend considering neuromonitoring for all surgical procedures in patients with SD and "spine-at-risk" [4]. They further state that neuromonitoring should be considered for patients with a "spine-at-risk" if the surgical procedure will involve an anesthetic exceeding 1 h. This recommendation was part of the practice guidelines put forth by White et al., and it includes all patients with SD [6]. Wake Up Safe, a subcommittee of the Society for Pediatric Anesthesia, has put forth a special advisory for patients with mucopolysaccharidosis (MPS) receiving general anesthesia [27]. Patients in this population have sustained neurological injury during anesthesia which was not detected until the patients were awake. All such patients had pre-existing kyphotic deformity of the spine and had multilevel spinal compromise and neural deficits. Thus, Wake Up Safe recommends considering neuromonitoring in MPS patients with kyphosis who are scheduled for a prolonged procedure. These patients should be positioned carefully, and spinal cord perfusion pressure should be preserved by maintaining mean arterial pressure near or above baseline. In addition, neuraxial blocks should be avoided for fear that local anesthetic (bolus dose with or without continuous infusion) injected may alter the neuraxial pressure dynamics in addition to contributing to hypotension; all of which compromise spinal cord perfusion [27].

18.2 Intraoperative Neuromonitoring

Intraoperative neuromonitoring involves the use of multiple modalities to evaluate the nervous system during surgical procedures where injury is possible. Multimodality testing has become the standard used by the IONM field because no one modality can capture the entire spectrum of information offered by the electrical activity of the neural structures. The main modalities are SSEPs, MEPs, EEG, EMG, and BAER. Somatosensory evoked potentials monitor sensory function, MEPs monitor motor function, EEG monitors cerebral cortex perfusion, EMG monitors the peripheral nerves, and BAER monitors the auditory pathway. Neuromonitoring is the use of electrical stimulation and the observed response. The characteristic waveform morphology, biphasic or polyphasic, is graphed as voltage over time and assessed to determine spinal cord and nerve integrity. Time from site of stimulation to arrival at recording site is latency (milliseconds). How strongly the stimulus is perceived is amplitude (microvolts), and it is measured as the peak height of the response. EP efficacy relies on the ability to detect lack of conduction or a decrease in evoked response to simulation. When there is decreased or no response to stimulation, neurologic compromise should be suspected. Changes in IONM can be multifactorial: curve correction, injury to the cord or nerves, ischemia, patient position, anesthetic medications, physiologic parameters, or electrode malfunction.

18.2.1 Somatosensory Evoked Potentials

Somatosensory evoked potentials evaluate the ascending sensory tracts; they monitor the integrity of the dorsal column of the spinal cord through which pass signals for fine touch, two-point discrimination, vibration, and proprioception. SSEPs summate the neural response of an electrical stimulation of a peripheral nerve: the posterior tibial or the peroneal in the lower extremity (LE) and the ulnar and median in the upper extremity (UE). The ulnar and median nerves are used from the distal UE and cover the cervical spine levels, from C5 down to T1. They also act as a control for the LE SSEP. The ulnar nerve is more frequently used because it travels all the way down to the T1 level and is more susceptible to positional and stretch injuries in the prone position, which is used for spine surgery. Therefore, a change in the ulnar SSEP may indicate a brachial plexus problem as well as a cervical spine issue. SSEP has the added benefit of avoiding placement too close to the median nerve and radial artery. The median nerve is more likely to be used if the C6 nerve root is specifically at risk. The fibular (peroneal) and posterior tibial nerves are used for the LE. The posterior tibial nerve is a larger response and therefore more reliable. It traverses the length of the spine from L5 through to the cervical spine and up to the brainstem and somatosensory cortex of the brain. Since it enters the spinal cord at L5, it does not cover the sacral roots or the sacral portion (conus) of the spinal cord. Thus, it would not be useful to monitor a decompression of the conus or to de-tether a filum.

The stimulus from upper or lower extremities travels in both directions but is recorded from the one traveling upwards. The stimulus travels from the peripheral nerve to the dorsal root ganglion and ascends the dorsal column spinal pathways to the lower medulla in the brainstem where it crosses midline and ascends in the medial lemniscus pathway. The signal then travels to the thalamus and provides sensory input to the sensorimotor cortex. The response may be recorded at any point along the pathway, which is advantageous to placement of an electrode. This may be from either a subcortical or cortical location. The subcortical location may be anywhere below the cortex, such as the popliteal fossa or the iliac crest for the LEs and the antecubital fossa or Erb's point for the UEs. Ultimately, both signals pass through the cervical spine to the brainstem, where they may be recorded by electrodes overlying the cervical vertebrae at the C2 or C5 level, and finally ascend to the somatosensory cortex. At the cortical level, standard EEG landmarks are used. The vertex of the skull is designated Cz, and C3 and C4 are 4 cm lateral to Cz on the left and right, respectively. SSEP of the lower extremities is recorded at Cz', which is 1–2 cm posterior to Cz. SSEP of the upper extremities is recorded at C3' and C4', which are 1–2 cm posterior to C3 and C4, respectively.

During spine surgeries, the subcortical and cortical locations are used because they are not in the surgical field. An amplitude reduction of more than 50% and a latency increase of greater than or equal to 10% should be investigated. SSEP specificity is 99.8% and sensitivity is 95% for monitoring of adolescent idiopathic scoliosis (AIS) [28]. This percentage changes when considering scoliosis other than AIS. Other published studies would place the sensitivity of the SSEP much lower—at less than 60% [29]. For these reasons, SSEPs always are paired with MEPs in current practice and rarely are used alone. However, they continue to provide useful information pertaining to prognosis and extent of an injury.

18.2.2 Motor Evoked Potentials

Motor evoked potentials monitor the descending motor pathways; they evaluate the spinal cord motor tracts located in the anterior spinal cord, which are responsible for pain and temperature sensation. It is a functional test of the corticospinal tract. The transcranial motor evoked potential (TcMEP) stimulates the motor cortex at the scalp and records the response, most commonly, from target muscles of interest. However, it also may be recorded from the spinal cord at various levels with epidural electrodes. Stimulation to either the scalp or to the spinal cord produces a muscle response, which is termed a compound muscle action potential (CMAP). While the CMAP is a very complex signal, it does not convey the entirety of electrical, physiological, and mechanical information inherent to a muscle contraction. The response from muscles requires a train of stimuli, not just a single stimulus, as the anterior horn cell will respond to a train but usually not to a single stimulus. The multi-pulse technique uses 5–9 electrical impulses of 300–700 mA with a pulse duration of 75 μs and 1000 p/s and a stimulation rate of 1 Hz. The adductor pollicis brevis is usually monitored in the UE and the tibialis anterior, lateral gastrocnemius, or abductor hallucis in the LE.

Subdermal needle electrodes stimulate the motor cortex, which transmits the stimulation through the internal capsule to the caudal medulla. Within the pyramidal tract, 80% of the fibers cross midline and descend into the lateral and anterior tracts of the spinal cord; the remaining 20% of the fibers stay ipsilateral. The motor tracts cover the anterior two-thirds of the spinal cord. These tracts then feed the anterior horn cell and the lower motor neurons through the spinal nerve roots. A direct wave (D-wave) due to direct excitation of the corticospinal tract neurons is followed by an indirect wave (I-wave) due to indirect depolarization and results from synaptic transmission of the stimulus. The D-wave is the orthodromic action potential resulting from direct white-matter stimulation with a single stimulus. The D-waves can only be recorded down to about the T9 or T10 cord level and are technically challenging. They also may receive interference from movement or instrumentation applied for deformity correction. Their use typically is restricted to spinal cord tumor excisions, for which they are sensitive. A 50% decline in amplitude constitutes the primary alert criterion, and a complete disappearance of the wave is highly correlated with significant neurologic deficit [30]. The I-waves are produced by the same stimulus, but they are produced by cortical neurons, which require synaptic activity. The I-wave activates α motor neurons with motor nerve transmission to the muscle. A CMAP is the result of muscle depolarization or direct stimulation of the nerve root or spinal cord. D-waves are less influenced by anesthetics as compared to I-waves because I-wave propagation involves a greater number of synapses. The I-waves require synaptic activity, and, therefore, they are inhibited by general anesthesia with inhalational agents. Along its entire length, this pathway is a very sensitive indicator of spinal cord contusion, compression, or ischemia.

The corticobulbar tract also is a motor pathway, which descends from the motor cortex and controls voluntary cranial and facial muscle movements. These pathways also may be monitored with MEP, especially for the lower face and tongue where a deficit would cause contralateral weakness of those muscles. There are various other pathways that, while not monitored directly, may affect the muscle MEPs by impacting muscle tone and postural control.

Motor evoked potential sensitivity and specificity for detecting corticospinal tract abnormality ranges from 65% to 100% [31, 32]. Alert criteria range between a 75% and 80% amplitude attenuation for corticospinal tracts. MEPs have a sensitivity of 100% and a specificity of 97.9% for peripheral nerve monitoring [33]. Alert criteria range from 50% to 65% amplitude attenuation for peripheral nerve alerts as compared to the 75–80% amplitude reduction criteria used for spinal cord alerts. Other criteria include the intermittent ability to record or a change in EP threshold if not related to anesthetic effects. An increase in threshold of 100 V or more, which persists for 1 h or more and is not related to systemic factors, correlates with potential postoperative motor deficit [34].

18.2.3 Electromyography

Electromyography (EMG) is an adjunct to SSEPs and MEPs and is used to record the electrical activity from motor units in the muscle. The two components of an EMG, spontaneous (sEMG) and triggered (tEMG), record muscle activity in response to stimulation of a nerve. sEMG is a passive recording modality in that electrodes are placed either in a muscle with a subdermal needle electrode or on the skin overlying a muscle with a cutaneous surface electrode. In an anesthetized patient, the muscle activity is limited, and the sEMG should be quiet at baseline. Neurotonic or EMG activity can indicate irritation of a nerve due to stretch, compression, or thermal activation. These parameters do not require comparison to other populations and are adequate for interpretation of individual patients. It is acknowledged as useful information about the muscles even though there are no normal parameters which are valid for this modality. Instead, it is understood that the raw data collected on an individual, where no changes have been made to the basic EMG set up or local environmental conditions, are valid for that individual [35]. sEMG is useful for monitoring peripheral and spinal nerve decompressions and certain joint procedures which may involve stretch of a nerve around a joint. With tEMG, the nerve is actively stimulated with a handheld electrical stimulator and a response may then be recorded from another segment of a nerve or muscle. This stimulation provides normalized information, such as the location, activation threshold, and speed of nerve conduction. It is used to locate and confirm the nerve is normal and within expected parameters. Alert criteria are based on normal thresholds for a nerve that is not diseased.

Electromyography is specific for a particular nerve root and is unlike SSEP and MEP, which assess several nerve roots simultaneously. In the intraoperative setting, EMG precludes the use of neuromuscular blockade during at least portions of the surgery. The subdermal needles used for recording are not ideal for the high frequency neurotonic spikes that may be a harbinger of nerve irritation. They also might not respond to a slow stretch, which is often the initiator of injury in long duration cases. An advantage to EMG is that it is not influenced by sedative or inhaled anesthetics [36].

18.2.4 Brainstem Auditory Evoked Potentials

Also known as auditory brainstem response, brainstem auditory evoked potentials (BAER) uses an audible "click" stimulus through earphones to stimulate the eighth cranial nerve (auditory nerve) and records an EP, which usually has five waves. Response for each wave correlates to a different segment along the auditory pathway. This EP is mostly used for acoustic neuroma excisions, microvascular decompressions, or cerebellar-pontine angle tumors. In the SD population, this response is used during foramen magnum, C1 decompressions, or high cervical instrumentation. This modality is preferred because it is sensitive to brainstem and cortical ischemia due to vertebral artery insufficiency. The response itself is impervious to halogenated anesthesia or neuromuscular blockade; however, for patients with SD who have a spine procedure, this modality is paired with MEPs and SSEPs, so the anesthetic requirements must be tailored to those modalities. Alerts are called when the amplitudes have a 50% or greater attenuation in waves IV and V or a 1-ms or more increase in the latency of waves IV and V. An increase in latency of 0.5 ms would result in a notification to the surgeon to check for, identify, and change any untoward factor as the signal change is approaching alert territory.

18.2.5 Electroencephalography

Electroencephalography (EEG) monitors the spontaneous activity of cortical neurons near recording electrodes and is a measure of electrical activity in different parts of the brain. EEG of the brain depends on metabolite and oxygen delivery. The EEG is used as an adjunct in the anesthesia depth evaluation and is beneficial in interpreting amplitude changes of EP signals—in particular, the MEP from target muscles or the CMAP. Cortical ischemia and seizure activity can be detected by EEG changes.

In SD cases, a modified-EEG set up typically is used consisting of two to four channels, primarily over the frontal lobe at Fp1 and Fp2 referenced to the parietal lobe at C3' and C4', or to FpZ. The EEG activity is correlated with anesthesia dosing, spontaneous EMG, heart rate, and mean arterial pressure. It also is useful for confirming that a patient's auto regulation is normal, as a drop in mean arterial pressure would be offset by vasoconstriction that feed the spinal cord, thus preserving EP amplitudes.

In cases where EEG is used to monitor cerebral perfusion, the alert criteria would be a slowing of EEG and an amplitude loss in the affected cortical areas. For cases involving the use of EEG to monitor anesthesia depth, there are no alert criteria. The EEG is used to assess the cortical suppression level and its correlation with the anesthesia depth.

18.2.6 Evoked Potential Variability

There is inherent variability with evoked potential (EP) monitoring that may result in fluctuating signal quality. MEPs are complicated by "noise," which is any signal interference or unwanted modification. Internal or inherent noise occurs throughout the entire sensory motor circuit (to and from the brain) through the spinal cord, in the soma and dendrites, through the anterior horn cells up and down the peripheral nervous system, to where it finally activates and responds to sensory neurons or muscle fibers. The random opening and closing of voltage or ligand gated ion channels, synaptic noise, synaptic variability, motor noise from muscle contraction, or neuronal cross talk generate electrical currents, causing noise [37]. The signals are inherently variable and are non-linear, meaning a constant stimulus does not result in a constant output but a variable output. SSEPs also are prone to variability due to ion channel fluctuations, synaptic noise, and cross talk variability. However, the SSEP response can be averaged, because the pathway is fixed; averaging provides a method to handle variability. Each individual stimulus is averaged by superimposing it on the prior average, producing a smoother wave form. MEPs cannot be averaged.

Accurate recordings also are dependent on many external factors, which introduce noise and signal integrity degradation. Factors that can affect recording accuracy include testing equipment capabilities, the user, active and reference electrode type and placement, signal transformation by amplifiers and filters, recording parameters like stimulus type and intensities, pulse durations, and the number and pattern of stimulus trains. Physiologic factors that impact the waveform amplitudes, latencies, and overall stability of the signals include patient positioning, anesthetic agents and anesthesia depth, temperature, hemoglobin, mean arterial pressure, and surgery duration. These are in addition to any co-existing medical problems, both known and unknown, the patient may have.

18.2.7 Stagnara Wake-Up Test

The Stagnara wake-up test is still the gold standard for intraoperative neurologic status assessment. If neurologic injury is suspected, the anesthesia level is reduced during the procedure and the patient is asked to move a certain body part on command. Usually, the hand grip is tested to confirm UE function, and foot dorsiflexion or toe movement tests LE function. The wake-up also allows for assessment of cognitive awareness. Surgery disruption, injury to the patient, or patient inability to cooperate due to physical or mental limitations make the wake-up a difficult tool. In addition, it tests only one point in time and is stressful for the patient; however, it continues to have value as an adjunct. The wake-up test can be used when the IONM information is equivocal, such as an unresolvable equipment issue or the drifting of EP signals close to the noise threshold as to be unreliable. Currently, IONM has become widely accepted as a continuous assessment tool for the integrity of the spinal cord and has largely replaced the Stagnara wake-up test [9, 24, 38].

18.3 Physiology and Risk of Neuromonitoring

18.3.1 Blood Supply to the Spinal Cord

The spinal cord receives dual blood supply from the aorta and its branches. Most of the blood supply is from the vertebral arteries, which give rise to two posterior spinal arteries (PSA) and one ASA. Vertebral artery anatomy has been reported to have the same degree of variability in the pediatric SD population compared to pediatric patients without spinal abnormality [39]. The ASA supplies the anterior two-thirds to four-fifths of the anterior spinal cord and the motor tracts. The PSAs supply the posterior one-third of the spinal cord and the sensory tracts. Collateral blood flow arises from cervical, intercostal, thoracic, and lumbar arteries. The PSAs receive more collateral blood flow than the ASA. The fewest radicular arteries are within the motor distribution of the thoracic spinal cord, and this is thus considered at highest risk for ischemia during hypotension, stretching, compressing, and corrective maneuvers. Since motor and sensory components of the spinal cord are separated, and because of the difference in their blood supply, SSEPs and MEPs are used simultaneously to provide a global status of the spinal cord. Also, because of the limited collateral flow to the anterior spinal cord, MEPs are more sensitive to ischemia and are considered an earlier predictor of impending damage to the cord.

18.3.2 Risk of Neuromonitoring

Neuromonitoring is a safe technique; however, there are complications that may result. Bleeding and infection from electrode insertion; arrythmia (0.03%); burns (0.05%); bite injuries of lips, tongue, or buccal membranes (0.2%); as well as endotracheal tube rupture have all been reported [40–42]. The incidence of seizures due to brain stimulation with IONM ranges from 0.7% to 5.4%. The highest reported incidence was associated with direct cortical stimulation combined with transcranial electrical stimulation [43]. However, seizures are not considered a contraindication for IONM [31]. Extremely rare complications, including compartment syndrome attributed to neuromonitoring needles and mandibular fracture due to masseter muscle stimulation, have been reported [44, 45]. In 2009, the Scoliosis Research Society favored multimodal monitoring despite the risks involved, most of which are rare [46].

18.4 Anesthesia

The impact of anesthetics on IONM may occur at multiple sites. The major effect is depression of synaptic function—the greater the number of synapses, the greater the effect of anesthetic agents [47, 48]. Additional effects from anesthetic agents may be due to depression of the reticular activating system, resulting in unconsciousness and lack of movement. In addition, some anesthetic agents may block synaptic function at the neuromuscular junction while others can cause changes in cardiovascular function [49].

Motor evoked and cortical potentials are more sensitive than SSEPs. Because most drugs produce a dose-dependent depression of the EEG, a decrease in amplitude and increase in latency is seen with cortical SSEPs. MEPs are susceptible to anesthetics at three sites: the motor cortex, the anterior horn cells, and the neuromuscular junction. The motor pathways and MEPs are moderately susceptible to anesthetics at the motor cortex, severely affected at the anterior horn cell, and only affected at the neuromuscular junction if neuromuscular blockade is used. The D-waves are unaffected by anesthetics because no synapses are involved in their production. However, I-waves, because they arise from interneuronal communication and require synaptic activity for their production, are more affected by anesthetics. The mechanics of depression vary depending on the anesthetic agent. For patients with SD, IONM mandates the anesthetic requirement be tailored to facilitate the most sensitive modality.

18.4.1 Inhaled and Intravenous Anesthetics

Somatosensory evoked potentials, MEPs, EMG, and EEG activity all are sensitive to both intravenous (IV) and inhalational anesthetic agents. The effects are dependent on dose and mechanism of action, specifically at the neurosynaptic and axonal pathways. Electrophysiological responses that depend on synaptic pathways are most susceptible to anesthetic agents. Longer pathways, like cortical-generated responses, are more affected by anesthetics than peripheral pathways (brainstem or spinal cord). Both inhalation and IV agents decrease waveform amplitudes and increase signal latency. Inhalational agents have greater effect and are more likely to suppress signal recording during IONM. Total intravenous anesthetics (TIVA) have proven to be more dependable with optimal signal attainment for patients undergoing IONM [50].

All halogenated inhalational anesthetics produce a change in the receptor or in the receptor ion channels [48, 51, 52]. The volatile anesthetics depress synaptic transmission in the anterior horn cells as well as communication between neurons, resulting in greater impact on the MEPs than SSEPs. With TcMEP, the suppressive effects of the inhalation agents can partially be compensated for by high voltage, multi-pulse transcranial stimulation [53]. SSEPs can be recorded at 0.5–1.0 MAC of inhalational agents. MEPs are more sensitive to the volatile anesthetics and are usually able to be recorded at 0.5 MAC. For patients with baseline neurological pathology, low concentrations of inhaled anesthetics may abolish the ability to obtain potentials [54]. Any factor which decreases signal quality, such as neurologic pathology, may contribute to neuromonitoring challenges. Nitrous oxide, which is an inhaled, but non-volatile anesthetic, reduces cortical SSEP amplitude and increases latency alone or when combined with volatile anesthetics or opioids. This is believed to be the result of blocking n-methyl-d-aspartate (NMDA) and acetylcholine receptors and activating opioid receptors.

18.4.2 Intravenous Anesthetics

Several intravenous agents bind to gamma-aminobutyric acid (GABA) receptors and increase chloride conduction, resulting in hyperpolarization of the cell membrane causing synaptic inhibition. Other agents block glutamic acid. Cortical SSEPs are minimally affected by IV agents, except for etomidate and ketamine. Most, not all, IV anesthetic agents cause a decrease in EEG amplitude and frequency.

A G-protein mechanism in opioids depress electrical excitability, resulting in increased inward potassium and decreased outward sodium current. They produce minimal effect on spinal- or sub-cortically recorded SSEPs and result in better SSEP responses from the decreased EMG activity. Cortical responses display mild depression of amplitude and increased latency. MEPs also have a minimal decrease in amplitude and increased latency [51, 55, 56]. Large doses of opioids, which result in EEG burst suppression, may affect MEPs. MEPs have the greatest reduction with sufentanil, followed by incrementally decreased reduction levels for alfentanil and fentanyl. The least suppressive narcotic is remifentanil, and the muscle MEPs are preserved at infusion rates of 0.6 μg/kg/min [53, 56]. Minimal opioid effects are likely due to minimal effect on the GABA and NMDA receptors and greater activity on the opioid receptor.

Barbiturates affect both motor and sensory potentials. The dose and specific barbiturate administered produce a variable response to the individual monitoring modality. Thiopental (TPL) decreases amplitude and increases latency transiently; whereas, phenobarbital does not affect SSEP. TPL and methohexital causes a dose-dependent decrease in CMAP amplitude, and TPL causes an increase in stimulation threshold. Methohexital can increase EEG activity [53].

Benzodiazepines act on GABA receptors, and administration of a benzodiazepine may cause mild depression of cortical SSEP amplitude with minimal change in latency [51, 56]. MEP amplitude and latency are minimally affected with diazepam at premedication doses. Midazolam, as a bolus or continuous infusion, has no significant effect with multi-pulse stimulation [51, 53, 56]. EEG slowing may occur with benzodiazepines [57].

Ketamine and etomidate can increase as well as depress potentials. Ketamine NMDA receptor activity inhibits nicotinic Ach receptors, decreases the presynaptic release of glutamate, and has opioid-like action on the opioid receptors. This results in enhanced cortical SSEP and MEP amplitude. EEG amplitude and frequency increases with ketamine.

Etomidate inhibits GABA, resulting in an increase in amplitude of cortical SSEP and decreases in MEP amplitude with no change in latency. This is believed to be due to disinhibition of the extrapyramidal system at the brain stem and spinal cord, resulting in increased excitability of the motor system at the spinal and supraspinal levels [53].

Propofol suppresses the activation of the alpha motor neurons at the spinal gray matter as well as enhances the inhibitory neurotransmitter, GABA, and causes synaptic inhibition [55]. It increases latency and decreases amplitude of cortical SSEP and TcMEPs [58, 59]. However, compared to the inhaled anesthetics, propofol results in better preservation of the cortical SSEP and MEP signal [53, 60]. Higher propofol infusion doses result in a greater number of stimuli needed to elicit a CMAP [56]. While propofol does cause structural changes in the waveform, the changes are predictable and stable.

Dexmedetomidine, an alpha two adrenergic agonist, has been used successfully as part of a TIVA without affecting monitoring [61]. SSEPs are not affected by dexmedetomidine; however, there are case studies that report decrease or loss of MEP with and without bolus dosing [62]. Dexmedetomidine has a half-life of 2 h and may cause a loss of the MEP, which can last several hours [63]. Others have reported dexmedetomidine at 0.5 μg/kg loading dose and 0.5 μg/kg/h infusion added to a propofol-remifentanil anesthetic did not attenuate the MEP and SSEP response in adult patients having thoracic spinal cord tumor resection [64]. Several animal studies have demonstrated dexmedetomidine is neuroprotective against spinal cord ischemia-reperfusion injury through suppression of spinal cord inflammation and neuronal apoptosis [65, 66]. A meta-analysis found both animal and human studies support a potential anti-inflammatory effect from dexmedetomidine [67]. Several more studies are needed on the effects of dexmedetomidine on neuromonitoring.

Neuromuscular blocking agents (NMBA) remove competing EMG activity to improve the SSEP waveform; however, MEPs, depending on degree of motor blockade, may be affected [68]. NMBAs bind at n-type acetylcholine receptors within the neuromuscular junction and can prevent recording of CMAP. Partial neuromuscular blockade allows recording of CMAP, sEMG, and tEMG. This requires careful titration of neuromuscular blockade to a train of four or comparison of CMAP amplitude before and after neuromuscular blockade [53].

18.5 Non-anesthetic Intraoperative Influences

Physiological factors also can impact IONM. Spinal cord blood flow, body temperature, oxygenation, ventilation, and technical issues may impact the EP signal.

Studies have shown cortical SSEPs are not able to be recorded at 12–15 mL/min/100 g cerebral blood flow [49]. In addition, SSEP changes have been seen with blood pressure that would not normally be seen with neural ischemia systolic blood pressure (SBP) greater than 90 mmHg. This, along with operative spinal distraction or stretch, vascular compromise from positioning, or ligation of perfusing arteries, or direct spinal cord injury, may result in greater effect than predicted by blood pressure alone. Patient positioning can cause neurological and vascular compromise and effect neuromonitoring as well [13]. Decreased blood flow to the cord has less effect on subcortical responses compared with cortical responses. In addition, MEPs are more sensitive to ischemia than the SSEPs. Responses may return to normal with increases in blood pressure. Increased intracranial pressure (ICP) decreases amplitude and increases latency of cortical SSEP. MEP latency increases with increasing ICP until there is no response.

A decrease in oxygen delivery and nutrients as seen with a hematocrit of 10–15% or a partial pressure (pCO_2) greater than 100 mmHg decreases SSEP and MEP amplitude and prolongs latency [68, 69]. Hypocapnia, pCO_2 less than 20 mmHg, causes cerebral vasoconstriction leading to ischemia and causes changes in MEP and SSEP [70].

Temperature changes also affect SSEP and MEP. Studies have demonstrated that hypothermia has variable effect on amplitude and causes an increase in latency and stimulation threshold [53, 71]. This is due to a decrease in neurotransmitter release, nerve depolarization and conduction velocity, and a decrease in sensory motor response to stimulation. MEP waveforms can be detected to core T of 31–34 °C with increasing latency stimulation threshold below 32 °C. Below 25 °C, SSEP and MEP disappear [72]. Hyperthermia decreases latency and amplitude of MEP. SSEP latency decreases up to 39 °C with no change in amplitude, and at temperatures greater than 39 °C, SSEP latency is prolonged [73]. At temperatures greater than 42 °C, cortical SSEP disappears, and spinal MEP deteriorates. At 45 °C, MEP amplitude decrease is irreversible [72, 74].

Evoked potential monitoring relies on synaptic transmission of electrical stimuli to provide data regarding the integrity of the spinal cord. Any physical parameter which changes impulse conduction along an axon may change the EP waveform. Patients of a younger age demonstrate an association with increase in threshold voltage needed to obtain MEP responses [75, 76]. This may be attributed to immaturity of motor pathways since myelination and synaptogenesis begin in utero and continue into the second decade [77]. Anesthetic agents used for TIVA have been successfully used for maintenance to obtain reliable potentials intraoperatively for children younger than 3 years old [77, 78].

18.6 Clinical Applications

Patients with SD benefit from multimodality IONM. Their pre-existing skeletal and neural pathology along with other comorbidities makes them high risk for surgery and anesthesia. The anesthetic for a patient with SD must be tailored carefully to maximize patient safety. The "spine-at-risk" patient may be identified by alerts in the electronic medical record based on diagnosis or the primary care team of physicians. Often the orthopedic or neurosurgical team makes the ultimate decision as to the need for neuromonitoring [79]. Patient care begins at a multidisciplinary weekly meeting. There, each patient's medical and surgical history is discussed and the need for neuromonitoring is established during this meeting.

When using IONM, it is essential that the choice of anesthetic consider the need for neuromonitoring to minimize anesthesia-related influences. Anesthetic management begins with careful patient positioning; this requires alignment of the mastoid process with the clavicle for a neutral position of the neck. After standard ASA monitors are applied, choice of anesthesia induction is influenced by patient age, anxiety, difficult airway history, and neurologic compromise. Agents used for anesthesia maintenance are driven by the anesthesiologist's discretion in consultation with the neurophysiologist monitoring the patient.

18.6.1 Regional Anesthesia and IONM

Regional anesthesia, if desired, may be performed in a patient who requires IONM. The only caveat is that the use of local anesthetic should be reserved for the end of the procedure when the IONM is concluded [80].

18.7 Case Illustration

18.7.1 Case 1

A 12-year-old male with McCune-Albright syndrome and known kyphoscoliosis had progressive, bilateral, LE weakness after sedation for a magnetic resonance image (MRI) of the spine. At the beginning of sedation, he had mean arterial pressures of 20s–50s, which lasted approximately 10 min; the rest of the sedation was uneventful. His MRI findings demonstrated narrowing of the spinal canal and cord compression. The anesthesia team was unaware of the findings and the patient was discharged home. Hours later he was readmitted to the hospital due to his inability to walk. An emergent spinal cord decompression was performed roughly 18 h following his MRI. He did regain most of his motor function after a month of physiotherapy. Since that time, patients who have a "spine-at-risk" and are scheduled for an MRI are initially positioned supine. Once the localizing sequence of the MRI is performed, the radiologist reviews it for spinal cord compression or edema. If there is cord compression or edema, the anesthesia team and the ordering surgeon are informed, the patient is positioned laterally, the length of the study is abbreviated, and mean arterial pressures are carefully monitored and maintained at baseline or higher, using vasopressors if needed.

18.7.2 Case 2

A 22-month-old female with spondyloepiphyseal dysplasia and foramen magnum stenosis with progressive atlanto-axial instability, presented for posterior cervical spine fusion at C1–C2, cervical decompression, and instrumentation with iliac bone graft and application of a halo. Induction, intubation, and halo application were uneventful. In addition, she had thoracolumbar kyphosis. Baseline MEP signals were obtained in the supine position. SSEP signals were not attempted because MEP is used to monitor any changes associated with positioning as it is quicker and more sensitive to instantaneous change. MEPs were present from the first dorsal interossei muscle in the UEs and the iliopsoas and abductor hallucis from the LEs. The tibialis anterior MEPs were not monitorable (Fig. 18.1). Upon turning the patient prone, neutral position of the head and neck was confirmed with imaging. At this point, MEPs were not present in the LEs, including the iliopsoas (Fig. 18.2). This made it unclear whether the loss of signals was a conduction block in the cervical spine or due to stenosis at the mid-thoracic level of the kyphosis. Adjustments to positioning were made by exchanging the hip roll for a smaller size and placing the roll upward to reduce traction. Anesthetic depth was assessed to be adequate and mean arterial pressure was noted to be greater than 65 mmHg. Positional effect on the spinal cord was suspected, and the patient was returned to the supine position upon which LE MEPs returned to the baseline range (Fig. 18.3). After a short period of time, the right-sided MEP signals were again diminished. At this point, the patient was turned to the left lateral position simultaneously removing the halo. Approximately 50 min later, MEPs were noted to be at baseline (Fig. 18.4). The decision was made not to proceed with the cervical decompression as it was unclear what level of the spine was causing the signal attenuation. The patient was awoken

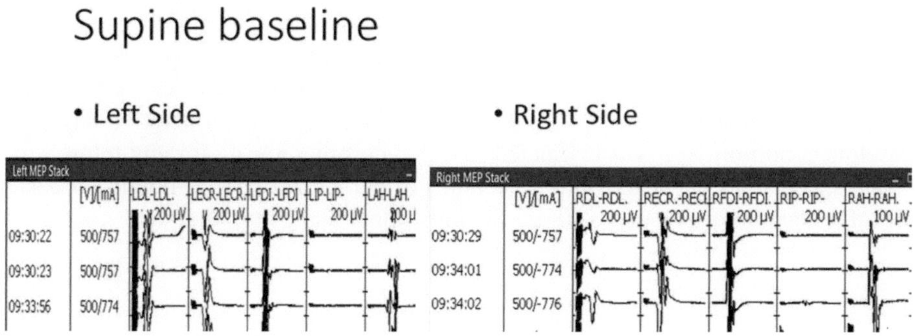

Fig. 18.1 Supine baseline evoked potentials: Cervical MEP present from upper extremities, no iliopsoas response is present; bilateral adductor hallucis potentials are present

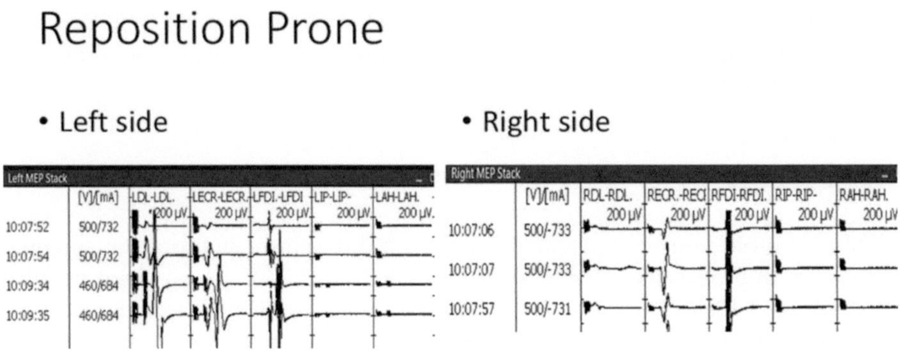

Fig. 18.2 Positioned in prone position: Right deltoid and bilateral adductor hallucis response are lost

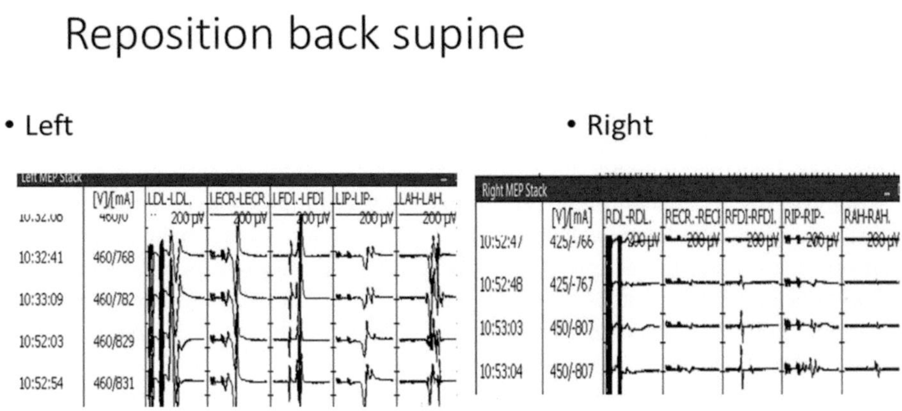

Fig. 18.3 After turning patient back to supine: Iliopsoas appeared, adductor hallucis returned, but right-side signals attenuated in both upper and lower extremities. Left side is improved

and able to spontaneously move all extremities. Immediately after waking, an MRI of the cervical/thoracic/lumbar spine without contrast was performed under general anesthesia. The MRI did not demonstrate any significant changes from the MRI conducted 6 months prior; it was speculated that the lack of signal change in the MRI was due to the period of insult (prone position) being brief enough that no irreversible damage occurred. The patient was discharged postoperative day 1 in baseline condition. She was instructed to have neuromonitoring for any future procedures, regardless of if they are procedures of her spine. After 4 months, she had T11–L1 decompression and T10–L3 fusion without any complications.

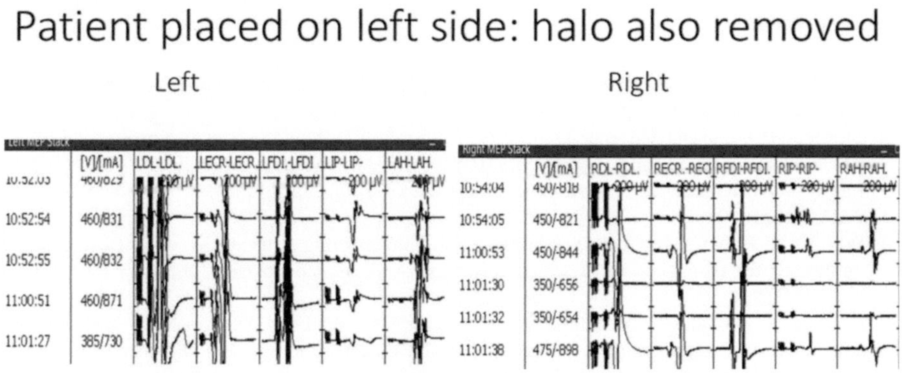

Fig. 18.4 Patient turned to left lateral position and halo removed: MEPs are now present from upper extremities, bilateral iliopsoas and bilateral adductor hallucis. Robust signals on left side and variable on the right side

18.8 Alerts

Managing IONM alerts requires a team approach. Alerts are based on a change to the baseline data collected on a particular patient.[1] Initially, the neurophysiologist assesses if there is a significant change in the signals, indicating a dysfunction of the corticospinal or dorsal column tracts. If there is a change, then the cause of the signal change must be addressed. In practical terms, this requires a systematic review of the potential sources of signal attenuation. Signal changes can be multifactorial and can be due to direct trauma to the cord, anesthetic agents, physiological changes, and technical issues with the equipment. Intervention to restore baseline signals may include radiologic studies, adding or removing medications, removing instrumentation, checking functionality of the equipment, or modifying the patient's position. Based on the intervention effects, the surgical team can decide to continue, change, terminate, or stage the procedure. There are published checklist recommendations for patients with a stable spine which may be used as a guideline for unstable spines [81].

Care of a patient with SD involves a thorough understanding of patient disease. IONM is used primarily to avoid iatrogenic neural injury during spinal fusion surgery. It also can provide information regarding a patient's electrophysiological profile, including the status of functional motor and somatosensory pathways. This requires a multidisciplinary approach and should be tailored to each patient. It is our position that all patients with SD presenting to the operating room for scoliosis surgery or those with a "spine-at-risk' requiring any surgery anticipated to last longer than 1 h should be monitored.

[1] Please refer to Chap. 7: Kniest Dysplasia for management of Alerts.

References

1. Cho SY, Jin DK. Guidelines for genetic skeletal dysplasias for pediatricians. Ann Pediatr Endocrinol Metab. 2015;20:187–91.
2. Mortier GR, Cohn DH, Cormier-Daire V, Hall C, Krakow D, Mundlos S, et al. Nosology and classification of genetic skeletal disorders: 2019 revision. Am J Med Genet A. 2019;179:2393–419.
3. Langston SJ, Krakow D, Chu A. Revisiting skeletal dysplasias in the newborn. NeoReviews. 2021;22:e216–29.
4. White KK, Bober MB, Cho TJ, Goldberg MJ, Hoover-Fong J, Irving M, et al. Best practice guidelines for management of spinal disorders in skeletal dysplasia. Orphanet J Rare Dis. 2020;15:161.
5. Dhiman N, Albaghdadi A, Zogg CK, Sharma M, Hoover-Fong JE, Ain MC, et al. Factors associated with health-related quality of life (HRQOL) in adults with short stature skeletal dysplasias. Qual Life Res. 2017;26:1337–48.
6. White KK, Bompadre V, Goldberg MJ, Bober MB, Cho TJ, Hoover-Fong JE, et al. Best practices in peri-operative management of patients with skeletal dysplasias. Am J Med Genet A. 2017;173:2584–95.
7. Upton AR, McComas AJ. The double crush in nerve entrapment syndromes. Lancet. 1973;2:359–62.
8. Nuwer MR, Dawson EG, Carlson LG, Kanim LE, Sherman JE. Somatosensory evoked potential spinal cord monitoring reduces neurologic deficits after scoliosis surgery: results of a large multicenter survey. Electroencephalogr Clin Neurophysiol. 1995;96:6–11.
9. Halsey MF, Myung KS, Ghag A, Vitale MG, Newton PO, de Kleuver M. Neurophysiological monitoring of spinal cord function during spinal deformity surgery: 2020 SRS neuromonitoring information statement. Spine Deform. 2020;8:591–6.
10. Berkowitz ID, Raja SN, Bender KS, Kopits SE. Dwarfs: pathophysiology and anesthetic implications. Anesthesiology. 1990;73:739–59.
11. Brill CB, Rose JS, Godmilow L, Sklower S, Willner J, Hirschhorn K. Spastic quadriparesis due to C1-C2 subluxation in Hurler syndrome. J Pediatr. 1978;92:441–3.
12. Beighton P, Craig J. Atlanto-axial subluxation in the Morquio syndrome. Report of a case. J Bone Joint Surg Br. 1973;55:478–81.
13. Appel S, Korn A, Biron T, Goldstein K, Rand N, Millgram M, et al. Efficacy of head repositioning in restoration of electrophysiological signals during cervical spine procedures. J Clin Neurophysiol. 2017;34:174–8.
14. Graham RB, Cotton M, Koht A, Koski TR. Loss of intraoperative neurological monitoring signals during flexed prone positioning on a hinged open frame during surgery for kyphoscoliosis correction: case report. J Neurosurg Spine. 2018;29:339–43.
15. Ofiram E, Lonstein JE, Skinner S, Perra JH. "The disappearing evoked potentials": a special problem of positioning patients with skeletal dysplasia: case report. Spine (Phila Pa 1976). 2006;31:E464–70.
16. Pruszczynski B, Mackenzie WG, Rogers K, White KK. Spinal cord injury after extremity surgery in children with thoracic kyphosis. Clin Orthop Relat Res. 2015;473:3315–20.
17. Drummond JC, Krane EJ, Tomatsu S, Theroux MC, Lee RR. Paraplegia after epidural-general anesthesia in a Morquio patient with moderate thoracic spinal stenosis. Can J Anaesth. 2015;62:45–9.
18. Hasan S, Arain A. Neuroanatomy, Spinal Cord Arteries. In: StatPearls [internet]. Treasure Island, FL: StatPearls Publishing; 2022. https://www.ncbi.nlm.nih.gov/books/NBK539889/.
19. Wang MX, Smith G, Albayram M. Spinal cord watershed infarction: novel findings on magnetic resonance imaging. Clin Imaging. 2019;55:71–5.
20. Colman MW, Hornicek FJ, Schwab JH. Spinal cord blood supply and its surgical implications. J Am Acad Orthop Surg. 2015;23:581–91.
21. Raggio CL, Yonko EA, Khan SI, Carter EM, Citron KP, Bostrom MPG, et al. Joint replacements in individuals with skeletal dysplasias: one institution's experience and response to operative complications. J Arthroplast. 2020;35:1993–2001.

22. Helenius IJ, Saarinen AJ, White KK, McClung A, Yazici M, Garg S, et al. Results of growth-friendly management of early-onset scoliosis in children with and without skeletal dysplasias: a matched comparison. Bone Joint J. 2019;101-B:1563–9.
23. Shafi K, Lovecchio F, Sava M, Steinhaus M, Samuel A, Carter E, et al. Complications and revisions after spine surgery in patients with skeletal dysplasia: have we improved? Global. Spine J. 2021;13(2):268–75.
24. Schwartz DM, Auerbach JD, Dormans JP, Flynn J, Drummond DS, Bowe JA, et al. Neurophysiological detection of impending spinal cord injury during scoliosis surgery. J Bone Joint Surg Am. 2007;89:2440–9.
25. Lyon R, Lieberman JA, Grabovac MT, Hu S. Strategies for managing decreased motor evoked potential signals while distracting the spine during correction of scoliosis. J Neurosurg Anesthesiol. 2004;16:167–70.
26. Pastorelli F, Di Silvestre M, Plasmati R, Michelucci R, Greggi T, Morigi A, et al. The prevention of neural complications in the surgical treatment of scoliosis: the role of the neurophysiological intraoperative monitoring. Eur Spine J. 2011;20(Suppl 1):S105–14.
27. Wake Up Safe. Spinal cord infarction or injury in patients with mucopolysaccharidoses (MPS) In: The Pediatric Anesthesia Quality Improvement Initiative. 2018. http://wakeupsafe.org/wp-content/uploads/2018/09/MPS_statement.pdf.
28. Thirumala PD, Bodily L, Tint D, Ward WT, Deeney VF, Crammond DJ, et al. Somatosensory-evoked potential monitoring during instrumented scoliosis corrective procedures: validity revisited. Spine J. 2014;14:1572–80.
29. Bhagat S, Durst A, Grover H, Blake J, Lutchman L, Rai AS, et al. An evaluation of multimodal spinal cord monitoring in scoliosis surgery: a single centre experience of 354 operations. Eur Spine J. 2015;24:1399–407.
30. Stecker MM. A review of intraoperative monitoring for spinal surgery. Surg Neurol Int. 2012;3:S174–87.
31. Wang S, Zhang J, Tian Y, Shen J, Zhao Y, Zhao H, et al. Intraoperative motor evoked potential monitoring to patients with preoperative spinal deficits: judging its feasibility and analyzing the significance of rapid signal loss. Spine J. 2017;17:777–83.
32. Miller SM, Donegan SW, Voigt N, Eltorai AEM, Nguyen J, Machan JT, et al. Transcranial motor-evoked potentials for prediction of postoperative neurologic and motor deficit following surgery for thoracolumbar scoliosis. Orthop Rev (Pavia). 2019;11:7757.
33. Wilent WB, Tesdahl EA, Harrop JS, Welch WC, Cannestra AF, Poelstra KA, et al. Utility of motor evoked potentials to diagnose and reduce lower extremity motor nerve root injuries during 4,386 extradural posterior lumbosacral spine procedures. Spine J. 2020;20:191–8.
34. Gonzalez AA, Jeyanandarajan D, Hansen C, Zada G, Hsieh PC. Intraoperative neurophysiological monitoring during spine surgery: a review. Neurosurg Focus. 2009;27:E6.
35. Halaki M, Ginn K. Normalization of EMG signals: to normalize or not to normalize and what to normalize to? In: Naik GR, editor. Computational intelligence in electromyography analysis: a perspective on current applications and future challenges. London: IntechOpen; 2012. p. 175–94.
36. Glover CD, Carling NP. Neuromonitoring for scoliosis surgery. Anesthesiol Clin. 2014;32:101–14.
37. Faisal AA, Selen LP, Wolpert DM. Noise in the nervous system. Nat Rev Neurosci. 2008;9:292–303.
38. Deletis V, Sala F. Intraoperative neurophysiological monitoring of the spinal cord during spinal cord and spine surgery: a review focus on the corticospinal tracts. Clin Neurophysiol. 2008;119:248–64.
39. Bauer JM, Dhillon ES, Kamps SE, Maloney E, Hsu M, Bompadre V, et al. Classifying vertebral artery anatomy abnormality in children with skeletal dysplasia. Spine Deform. 2021;9:833–9.
40. Legatt AD. Current practice of motor evoked potential monitoring: results of a survey. J Clin Neurophysiol. 2002;19:454–60.
41. Macdonald DB. Intraoperative motor evoked potential monitoring: overview and update. J Clin Monit Comput. 2006;20:347–77.

42. Sanders A, Andras L, Lehman A, Bridges N, Skaggs DL. Dermal discolorations and burns at neuromonitoring electrodes in pediatric spine surgery. Spine (Phila Pa 1976). 2017;42:20–4.
43. Ulkatan S, Jaramillo AM, Téllez MJ, Kim J, Deletis V, Seidel K. Incidence of intraoperative seizures during motor evoked potential monitoring in a large cohort of patients undergoing different surgical procedures. J Neurosurg. 2017;126:1296–302.
44. Chalkeidis O, Kalakonas A, Chaidutis C, Chotoumanidis C. Endotracheal tube cutting during neurosurgical operation. Eur J Anaesthesiol. 2009;26:533–4.
45. Eli IM, Gamboa NT, Guan J, Taussky P. Acute compartment syndrome as a complication of the use of intraoperative neuromonitoring needle electrodes. World Neurosurg. 2018;112:247–9.
46. Scoliosis Research Society. Neuromonitoring Information Statement. In: Quality & safety: position statements. 2019. https://www.srs.org/about-srs/quality-and-safety/position-statements/neuromonitoring-information-statement.
47. Deletis V. Intraoperative monitoring of the functional integrity of the motor pathways. Adv Neurol. 1993;63:201–14.
48. Sloan TB, Heyer EJ. Anesthesia for intraoperative neurophysiologic monitoring of the spinal cord. J Clin Neurophysiol. 2002;19:430–43.
49. Sloan T. Anesthesia and intraoperative neurophysiological monitoring in children. Childs Nerv Syst. 2010;26:227–35.
50. Pechstein U, Nadstawek J, Zentner J, Schramm J. Isoflurane plus nitrous oxide versus propofol for recording of motor evoked potentials after high frequency repetitive electrical stimulation. Electroencephalogr Clin Neurophysiol. 1998;108:175–81.
51. Kalkman CJ, Drummond JC, Ribberink AA. Low concentrations of isoflurane abolish motor evoked responses to transcranial electrical stimulation during nitrous oxide/opioid anesthesia in humans. Anesth Analg. 1991;73:410–5.
52. Zentner J, Albrecht T, Heuser D. Influence of halothane, enflurane, and isoflurane on motor evoked potentials. Neurosurgery. 1992;31:298–305.
53. Lotto ML, Banoub M, Schubert A. Effects of anesthetic agents and physiologic changes on intraoperative motor evoked potentials. J Neurosurg Anesthesiol. 2004;16:32–42.
54. Macdonald DB, Skinner S, Shils J, Yingling C, American Society of Neurophysiological Monitoring. Intraoperative motor evoked potential monitoring: a position statement by the American Society of Neurophysiological Monitoring. Clin Neurophysiol. 2013;124:2291–316.
55. Kakinohana M, Fuchigami T, Nakamura S, Kawabata T, Sugahara K. Propofol reduces spinal motor neuron excitability in humans. Anesth Analg. 2002;94:1586–8.
56. Scheufler KM, Zentner J. Motor-evoked potential facilitation during progressive cortical suppression by propofol. Anesth Analg. 2002;94:907–12.
57. Sloan TB, Fugina M, Toleikis JR. Effects of midazolam on median nerve somatosensory evoked potentials. Br J Anaesth. 1990;64:590–3.
58. Kawaguchi M, Sakamoto T, Inoue S, Kakimoto M, Furuya H, Morimoto T, et al. Low dose propofol as a supplement to ketamine-based anesthesia during intraoperative monitoring of motor-evoked potentials. Spine (Phila Pa 1976). 2000;25:974–9.
59. Nathan N, Tabaraud F, Lacroix F, Mouliès D, Viviand X, Lansade A, et al. Influence of propofol concentrations on multipulse transcranial motor evoked potentials. Br J Anaesth. 2003;91:493–7.
60. Deletis V, Kiprovski K, Morota N. The influence of halothane, enflurane, and isoflurane on motor evoked potentials. Neurosurgery. 1993;33:173–4.
61. Bala E, Sessler DI, Nair DR, McLain R, Dalton JE, Farag E. Motor and somatosensory evoked potentials are well maintained in patients given dexmedetomidine during spine surgery. Anesthesiology. 2008;109:417–25.
62. Tobias JD, Goble TJ, Bates G, Anderson JT, Hoernschemeyer DG. Effects of dexmedetomidine on intraoperative motor and somatosensory evoked potential monitoring during spinal surgery in adolescents. Paediatr Anaesth. 2008;18:1082–8.
63. Mahmoud M, Sadhasivam S, Salisbury S, Nick TG, Schnell B, Sestokas AK, et al. Susceptibility of transcranial electric motor-evoked potentials to varying targeted blood levels of dexmedetomidine during spine surgery. Anesthesiology. 2010;112:1364–73.

64. Li Y, Meng L, Peng Y, Qiao H, Guo L, Han R, et al. Effects of dexmedetomidine on motor- and somatosensory-evoked potentials in patients with thoracic spinal cord tumor: a randomized controlled trial. BMC Anesthesiol. 2016;16:51.
65. Bell MT, Puskas F, Smith PD, Agoston VA, Fullerton DA, Meng X, et al. Attenuation of spinal cord ischemia-reperfusion injury by specific α-2a receptor activation with dexmedetomidine. J Vasc Surg. 2012;56:1398–402.
66. Sun Z, Zhao T, Lv S, Gao Y, Masters J, Weng H. Dexmedetomidine attenuates spinal cord ischemia-reperfusion injury through both anti-inflammation and anti-apoptosis mechanisms in rabbits. J Transl Med. 2018;16:209.
67. Flanders CA, Rocke AS, Edwardson SA, Baillie JK, Walsh TS. The effect of dexmedetomidine and clonidine on the inflammatory response in critical illness: a systematic review of animal and human studies. Crit Care. 2019;23:402.
68. Pajewski TN, Arlet V, Phillips LH. Current approach on spinal cord monitoring: the point of view of the neurologist, the anesthesiologist and the spine surgeon. Eur Spine J. 2007;16(Suppl 2):S115–29.
69. Nagao S, Roccaforte P, Moody RA. The effects of isovolemic hemodilution and reinfusion of packed erythrocytes on somatosensory and visual evoked potentials. J Surg Res. 1978;25:530–7.
70. Ghatol D, Widrich J. Intraoperative neurophysiological monitoring. 2023 Jul 24. In: StatPearls [internet]. Treasure Island, FL: StatPearls Publishing; 2024.
71. Browning JL, Heizer ML, Baskin DS. Variations in corticomotor and somatosensory evoked potentials: effects of temperature, halothane anesthesia, and arterial partial pressure of CO_2. Anesth Analg. 1992;74:643–8.
72. Oro J, Haghighi SS. Effects of altering core body temperature on somatosensory and motor evoked potentials in rats. Spine (Phila Pa 1976). 1992;17:498–503.
73. Banoub M, Tetzlaff JE, Schubert A. Pharmacologic and physiologic influences affecting sensory evoked potentials: implications for perioperative monitoring. Anesthesiology. 2003;99:716–37.
74. Yamane T, Tateishi A, Cho S, Manabe S, Yamanashi M, Dezawa A, et al. The effects of hyperthermia on the spinal cord. Spine (Phila Pa 1976). 1992;17:1386–91.
75. Chen X, Sterio D, Ming X, Para DD, Butusova M, Tong T, et al. Success rate of motor evoked potentials for intraoperative neurophysiologic monitoring: effects of age, lesion location, and preoperative neurologic deficits. J Clin Neurophysiol. 2007;24:281–5.
76. Lieberman JA, Lyon R, Feiner J, Diab M, Gregory GA. The effect of age on motor evoked potentials in children under propofol/isoflurane anesthesia. Anesth Analg. 2006;103:316–21.
77. Fulkerson DH, Satyan KB, Wilder LM, Riviello JJ, Stayer SA, Whitehead WE, et al. Intraoperative monitoring of motor evoked potentials in very young children. J Neurosurg Pediatr. 2011;7:331–7.
78. Herta J, Yildiz E, Marhofer D, Czech T, Reinprecht A, Rössler K, et al. Feasibility of intraoperative motor evoked potential monitoring during tethered cord surgery in infants younger than 12 months. Childs Nerv Syst. 2022;38:397–405.
79. Kandil AI, Pettit CS, Berry LN, Busso VO, Careskey M, Chesnut E, et al. Tertiary pediatric academic institution's experience with intraoperative neuromonitoring for nonspinal surgery in children with mucopolysaccharidosis, based on a novel evidence-based care algorithm. Anesth Analg. 2020;130:1678–84.
80. Eiszner JR, Atanda A Jr, Rangavajjula A, Theroux M. A case series of peripheral nerve blocks in pediatrics and young adults with skeletal dysplasia. Paediatr Anaesth. 2016;26:553–6.
81. Vitale MG, Skaggs DL, Pace GI, Wright ML, Matsumoto H, Anderson RCE, et al. Best practices in intraoperative neuromonitoring in spine deformity surgery: development of an intraoperative checklist to optimize response. Spine Deform. 2014;2:333–9.

Part IV

Pain Management Including Regional Anesthesia

Regional Anesthesia and Pain Management

19

Robert S. Lang, Connie J. Lin, and Kesavan Sadacharam

19.1 Introduction

The practice of regional anesthesia in children has dramatically increased over the last decade due to the increased adoption of multi-modal analgesia to decrease perioperative opioid use and to lessen the postoperative prescription of opioids. In our practice, the use of regional anesthesia has increased from less than 10% to more than 30% of the overall surgical population over the last decade. This increase in the adoption of regional anesthesia reflects the general trend across the world [1].

19.2 Regional Anesthesia Improves Perioperative Patient Outcomes

Regional anesthesia as part of multi-modal analgesia has been shown to provide superior pain outcomes with fewer complications, such as nausea and vomiting, in comparison to opioid-based perioperative analgesia [2, 3]. The ability to prolong the block with perineural adjuncts or continuous peripheral nerve block (PNB) catheters has extended the block duration beyond the immediate postoperative period. In addition, including regional anesthesia in enhanced recovery protocols

R. S. Lang (✉) · C. J. Lin · K. Sadacharam
Department of Anesthesiology and Perioperative Medicine, Nemours Children's Hospital, Wilmington, Delaware, USA

Anesthesiology and Pediatrics, Sidney Kimmel Medical College, Thomas Jefferson University, Philadelphia, Pennsylvania, USA
e-mail: Robert.Lang@nemours.org; Connie.Lin@nemours.org; Kesavan.Sadacharam@nemours.org

M. C. Theroux (ed.), *Anesthesia Care of Patients with Skeletal Dysplasia*,
https://doi.org/10.1007/978-3-031-92501-6_19

for various surgeries has improved the recovery of the patients in terms of quality and speed [4, 5]. Traditionally, peripheral nerve blocks were performed using the landmark technique. However, over the last decade, using ultrasound to perform peripheral nerve blocks has become a standard of care in clinical practice. Ultrasound use facilitates the visualization of the anatomic structures and thus makes the blocks safer and more efficient. Visualizing the anatomy of the nerves and the spread of injected local anesthetic with ultrasound has been shown to decrease the complications related to blocks and improve the success rate of the peripheral nerve blocks in children [6, 7]. In addition, ultrasound has been shown to decrease the number of needle passes and the time to perform the block [7]. Thus, indirectly, it helps with the satisfaction of the anesthesiologists and the surgeons because the peripheral nerve blocks often are performed in children asleep in the operating room.

Apart from improved pain control, in recent times, there has been an increased awareness of the significance of patient or parental satisfaction in value-based health care. Poorly controlled pain is associated with perioperative anxiety for both children and their parents [8]. Thus, the optimal control of perioperative pain will provide comfort, a better perioperative experience, and improve parental and patient satisfaction. Over the last decade, there has been an increasing awareness of opioid use during the perioperative period and its contribution to overall opioid use disorders, both short and long terms. A small proportion of surgical patients who had opioids prescribed immediately after surgery continued to use them long after the surgery [9]. Opioid exposure during the surgery, along with other risk factors, such as multiple surgeries, may increase the risk of chronic opioid use in an otherwise opioid-naïve patient [10].

Patients with skeletal dysplasia (SD) are unique in that they need multiple surgeries during their lifetime. This increases exposure to opioids in comparison to the general population. In addition, numerous opioid exposures could lead to opioid sensitization, which may further lead to an increased need for opioids with subsequent surgeries. Thus, regional anesthesia, as part of multi-modal analgesia, plays a significant role in this patient population to minimize the exposure to opioids and subsequent risk of chronic opioid use. In addition, patients with skeletal dysplasia are known to have a higher risk of concomitant structural airway deformity and sleep apnea in comparison to general population [11, 12]. The administration of opioids during the perioperative period could exaggerate or increase the risk of airway-related complications such as respiratory depression and airway obstruction. Regional anesthesia as part of multi-modal analgesia in this patient population not only helps with the quality of pain control, but also the safety of recovery in the immediate postoperative period.

19.3 Challenges with Regional Anesthesia in Patients with Skeletal Dysplasia

Despite the growth of regional anesthesia in adults and children over the last decade, the practice and adoption of regional anesthesia in patients with SD has been gradual. The patients with skeletal dysplasia are medically complex and technically challenging in various ways, rendering regional anesthesia, both central and peripheral, more difficult to perform. In children, the most performed regional techniques are caudal or epidural anesthesia. Caudal and epidural blocks are traditional regional techniques with a long history of efficacy and safety. Additionally, they can be performed with a landmark technique with which most anesthesiologists are proficient. However, patients with SD are unique because of the high incidence of spinal deformities such as atlantoaxial instability, spinal cord stenosis, and kyphoscoliosis. The presence of a spinal deformity can make the performance of the caudal and epidural blocks in these patients challenging. In addition, these patients are at high risk of spinal cord injury during the surgery both because of positional compromise on the spinal cord as well as compromise of spinal cord perfusion. The latter may occur due to the use of an epidural anesthetic, which requires the injection of the local anesthetic in the narrow space around the spinal cord [13]. Case reports of such spinal cord injury list possible mechanistic factors, including the relative hypotension often observed when a neuraxial anesthetic such as epidural anesthesia is performed [14]. In our practice, we do not perform caudal or epidural blocks in patients with SD due to concern of spinal cord injury, both directly during the performance of the procedure and indirectly due to inal cord injury, both directly during the performance pressure-related injury because of injection of local anesthetic into the narrow epidural space.

In patients with SD, there are multiple challenges with performing peripheral nerve blocks. First, patients with SD are small for their age and have a lower weight compared to the general population. Most often, these children undergo complex extremity surgeries involving multiple dermatomes on both lower extremities. The combination of the low weight along with multi-dermatomal surgeries involving bilateral extremities makes it challenging to dose the local anesthetic for multiple blocks successfully. In addition, most often, these complex surgeries necessitate placement of a catheter or catheters, and low weight limits the dose of infusion of local anesthetics for postoperative pain control. Thus, these patients need special consideration and careful pre-procedure planning in terms of concentration and volume of local anesthetic to balance the safety and efficacy of the peripheral nerve blocks. The other significant challenge is the complexity of both gross and sonoanatomy of these patients. The extremities of patients in this population are often underdeveloped both because of the underlying disease and because of limitations with ambulation and range-of-motion in some instances. In addition, it is common for patients with SD to have contractures of the extremities. The underdevelopment of limbs and contracture of the extremities pose challenges to the successful performance of the PNB.

Children with SD most often have scoliosis or kyphosis with an unstable spine [12]. The combination of an unstable spine and the long duration of surgery places

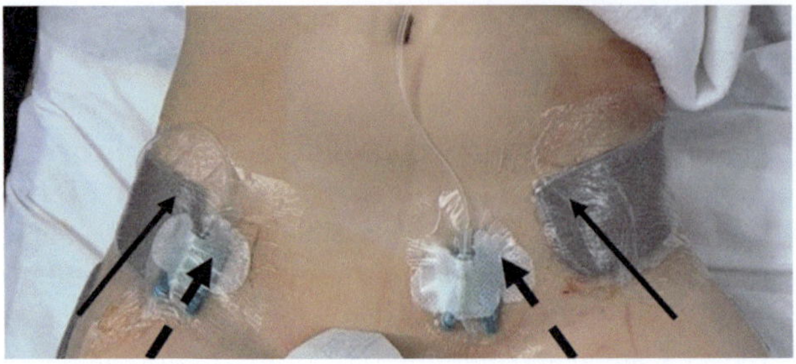

Fig. 19.1 Overlapping surgical dressings (solid arrows) with peripheral nerve (Fascia Iliaca) block catheters (dotted arrows)

them at risk for spinal cord injuries. So, in our practice, it is a standard of care to monitor the integrity of the spinal cord using neuromonitoring for lengthy surgeries. The use of neuromonitoring precludes the use of regional anesthesia, both the single shot and the continuous catheter before the surgery, as a preemptive analgesia. Thus, in most instances, at our practice, we perform the regional blocks at the end of the surgery for postoperative pain control and recovery. The performance of the PNB at the end of the surgery has its disadvantages, such as alteration of the anatomy due to the procedure and overlapping surgical dressings, wounds, and casts around the area of the block's performance (Figs. 19.1 and 19.2). As a result, sometimes we alter the type of PNB and procedural technique needed to perform a particular block. There may be instances when there is a need to consider a different approach to a particular block, such as femoral nerve block versus fascia iliaca block, or different visualization techniques, such as out-of-plane versus in-plane needle visualization.

19.4 Practice of Regional Anesthesia in Patients with Skeletal Dysplasia

In our practice, PNBs are selectively utilized for osteotomy procedures involving the hip, femur, tibia, or fibula. The most common type of skeletal dysplasia for lower extremity osteotomy surgery in our practice is Morquio syndrome, followed by spondyloepiphyseal dysplasia, achondroplasia, Kniest dysplasia, Hurler syndrome, and diastrophic dysplasia. Each of these types of skeletal dysplasia is unique and poses different challenges for anesthetic management in general,[1] performance of PNBs, and postoperative pain management. The kind of extremity osteotomy surgery performed on these patients is further influenced by various factors such as type of SD, diagnosis of deformity, age, and the functional needs of the patient. Femur derotational osteotomy is the most frequently performed osteotomy in this patient population, either on its own or, more often, in combination with either epiphysiodesis or hip osteotomies. Other commonly performed procedures include

[1] Please refer to the individual chapters for different dysplasias.

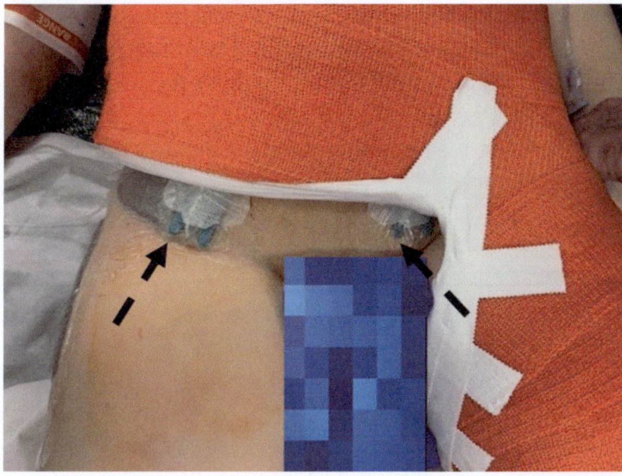

Fig. 19.2 Cast around the area of the block's performance and indwelling catheters (dotted arrows). Both catheters appear partially hidden by the surrounding cast but do not interfere with examining the catheter sites and eventual removal of the catheters

iliac osteotomies, hip replacement, and tibia and fibula osteotomies. The primary osteotomy influences the type of PNB, as the procedure often involves multiple dermatomes. Femoral nerve or fascia iliaca catheters are alternatively used for the femur and iliac osteotomies based on the preference of the anesthesiologists and ease of performance of the block at the end of the procedure due to overlap of either dressing or the wound. However, most of our patients had lumbar plexus catheters for the hip replacement.[2] We routinely place dry catheters before the surgery and bolus them with local anesthetics once the neuromonitoring is complete at the end of surgery. The patients who had tibia or fibula osteotomies receive sciatic catheters, most often at the popliteal level, and single-shot femoral or adductor canal blocks. Other approaches to the sciatic nerve utilized here include the anterior, subgluteal, and infragluteal approaches. Our practice is to start the infusion of local anesthetic in the recovery area and continue for 2 or 3 days after the surgery. The preferred local anesthetic in our practice is 0.2% ropivacaine, both for the bolus and infusion for postoperative pain management. We limit the postoperative infusion to 0.4 mg/kg/h. The dedicated pain team follows these patients for the duration of the infusion of the local anesthetic and further if there is a need for continued pain control. The catheter is removed in all patients before they are discharged home.

In summary, peripheral nerve blocks have provided us with a successful alternate option to the central neuraxial blockade. Recently, PNBs have become an integral part of multi-modal analgesia for perioperative pain management in this patient population for lower extremity osteotomy surgical procedures. In our experience, we found that PNBs have improved the quality of pain control after the surgery,

[2] Please refer to the Chap. 14 "Morquio A (MPS IVA) Syndrome" for additional caution advised for patients with MPS.

facilitated lower use of opioids, and thus contributed to lessening the opioid-related complications [15]. In addition, we also notice that the quality of recovery and parental and patient satisfaction have improved with the use of PNBs as part of the multi-modal analgesia. However, patients with SD present unique challenges to perform regional anesthesia in a safe and successful manner due to complex anatomy and other logistical challenges of the perioperative period that are distinctive for this patient population.

19.5 Challenges with Skeletal Dysplasia and Pain Management

Pain management for patients with SD is like that of most other patients. Maximizing therapeutic effectiveness while minimizing side effects holds true for all postoperative patients. A thorough evaluation of each patient with SD plays an integral role in the proper management of pain after surgical interventions. Pain is a multidimensional phenomenon that combines the patient's experiences and expectations as they interact with a noxious stimulus. Pain in each patient is unique, and management should be individualized as much as possible. In a cross-sectional assessment of pain by Alade et al., there was a higher percentage of pain prevalence in adult patients with SD than in other US adult populations [16]. This was highest in achondroplasia, diastrophic dysplasia, and hypochondroplasia [16]. Chronic pain in SD can include the lower back, extremities, and various joints with decreased measures of quality of life compared to the general population [16, 17]. Jennings et al. showed an association between pain and mental health conditions in this population. Understanding this and other factors related to skeletal dysplasia allows for proper assessment and treatment of pain [17]. Outlined in this section are considerations of certain challenges associated with the myriad SD syndromes and possible interventions to mitigate postoperative complications. Because over 400 SD syndromes are known, exact recommendations for each condition will not be presented, but rather guidelines to assist the practitioner in developing individualized treatments for their skeletal dysplasia patients [18].

19.6 Assessment

Pain control is well recognized as an important portion of the care of a postsurgical patient. Frequent patient pain assessment is incorporated into their care regardless of intervention or surgical procedure. Assessment should include objective and subjective measures to form a complete picture of the patient's experience. Pain assessment has been studied numerous times with the development of many tools to guide the practitioner. These tools help to develop a more complete picture of the patient's pain experience. Objective measures such as pain-related behaviors and vital signs are only a portion of the assessment. Measuring pain at rest and during physical therapy or other activities demonstrates how well the patient can rest and mobilize, which are vital aspects of recovery. Commonly used assessment tools include the Numeric Rating Scales (NRS), Visual Analog Scales (VAS), Faces Pain Scales (FPS), and many more [19]. There are pain assessments that can be used for patients without impairment, critically ill or sedated patients, or those with cognitive difficulties. Child-specific pain assessments are numerous, with several self-report scales, observer-related scales, and scales for cognitively impaired children [20]. Neonatal and infant scales are also available. These tools have varying degrees of efficacy for different patients, so choosing the proper evaluation tool is important for pain management.

In addition to objective and subjective measures of pain, understanding the patient's perspectives and expectations can help facilitate an effective relationship. Setting realistic goals for pain management, even before surgery, can help limit the overuse of medications. In many cases, complete pain relief is not possible, and the patient will need to understand this. Making the pain manageable enough to provide relief for effective rest and physical rehabilitation without unwanted side effects of pain medications should be the goal for all patients. A patient's pain experience may change as multiple subsequent procedures are performed. A patient may gain a perspective on their pain management that facilitates their involvement in proper management, or a patient may regress, making pain management difficult. Understanding and addressing the patient's mental health is essential to pain assessment. Effective pain management may also necessitate assessing and addressing anxiety or depressive moods. Interacting with other health care providers or consult services may be necessary.

19.7 Pre-existing Conditions

Many patients may have pre-existing pain syndromes. Surgical procedures may exacerbate specific pain syndromes; whereas, some procedures may be aimed at alleviating pain. Amplified pain syndrome and other chronic conditions may be present in SD patients. Specific forms of SD predispose the patient to cervical myelopathy and cervical, thoracic, or lumbar cord compression. Joint pain, back pain, head and neck pain, and muscular pain may all be present in a patient before surgical procedures. Pain has been noted to be a significant and underreported symptom of skeletal dysplasias such as Morquio A syndrome [21]. Many of these conditions not only predispose the patient to have pain but also may necessitate surgical interventions such as pelvic and femoral osteotomies, hip arthroplasties, and spinal decompressions [21]. The presence of these conditions and any preoperative analgesic use can make postoperative pain management more difficult. To develop an appropriate and effective analgesic plan, one must assess pre-existing pain and any current pain medications. Some patients may present with long-term opioid use, and assessing their baseline needs before surgery will help establish a starting point.

A significant consideration of the patient with SD is the presence of obstructive sleep apnea and the risk of a difficult airway. Airway management will typically be performed in a controlled environment, such as the operating room or other anesthesia induction suites, with a complement of tools at the manipulator's disposal. Time, consideration, and preparation go into successfully managing many SD patients. When a patient is recovering from surgery and requiring opioid medications, a risk of respiratory depression and loss of airway is present. Depending on the urgency, airway management on the hospital floor or ICUs may be more complex and less organized. In some cases, the difficulty of the airway may preclude successful management, leading to loss of oxygenation and ventilation with subsequent comorbidities or mortality. Minimizing the use of medications that may predispose a patient to respiratory depression is a key ideal when developing postsurgical pain strategies.

19.8 Management Strategies

Establishing appropriate pain expectations when introducing multi-modal practices may be difficult if the patient has become familiar with opioids and their use. Though opioids may be used in a limited amount or even unnecessary postoperatively when using multi-modal techniques, a previously established amount may be anticipated and expected by the patient. The use of other medications in a multimodal approach to pain management should be reviewed preoperatively to establish new expectations. These medications may be part of an established pain pathway or an enhanced recovery after surgery (ERAS) pathway. These pathways are often established to minimize opioids, which will maximize recovery while avoiding respiratory depression, nausea and vomiting, constipation, and more [22–24]. Early mobilization, enteral intake, and deep breathing exercises all assist in the recovery process. Potent analgesics, such as the NMDA antagonist ketamine, can be used to maintain respiratory drive and avoid many other side effects associated with opioids. Additionally, pathways may also include the use of acetaminophen; alpha-2 adrenergic agonists, such as clonidine or dexmedetomidine; non-steroidal anti-inflammatory medications, such as ketorolac anti-convulsants, like gabapentin and pregabalin; and local anesthetics in the form of lidocaine infusions, local infiltration, or regional anesthetics [24–26]. Surgical intervention and a thorough patient evaluation will dictate the use of these medications and their combination. Regional anesthetics are emerging as a highly effective method in managing postoperative pain while minimizing opioid use in SDs such as osteogenesis imperfecta [26]. Some of these techniques have been discussed earlier in this chapter.

Establishing trust with the patient and their family will help facilitate the exploration of alternative pain management techniques. Techniques can include different classes of pain medications, as stated above, and the development of coping strategies, distraction techniques, and breathing exercises [27]. The practitioner can also consider specific strategies that are currently being examined, such as massage, nerve stimulation technologies, acupuncture, and more [28, 29]. Techniques focused on managing anxiety and other mental health conditions can also play an important role in postoperative management [27].

With an understanding of the implications of SD on pain management and the utilization of multi-modal techniques, a safe and effective strategy can be developed for even the most complicated patients. Establishing a robust system to help these patients navigate the surgical procedure and recovery process while minimizing pain and medication side effects can foster a nurturing and healthy environment for healing. The literature on the proper management of SD patients continues to grow, and a thorough knowledge with thoughtful application of pain management strategies allows practitioners to excel in the care of their patients.

References

1. Dadure C, Veyckemans F, Bringuier S, Habre W. Epidemiology of regional anesthesia in children: lessons learned from the European Multi-Institutional Study APRICOT. Paediatr Anaesth. 2019;29:1128–35.
2. Arce Villalobos M, Veneziano G, Iobst C, Miller R, Walch AG, Roth C, et al. Regional anesthesia for pain management after orthopedic procedures for treatment of lower extremity length discrepancy. J Pain Res. 2020;13:547–52.
3. Muhly WT, Ganley T, Jantzen E, Browne P, Kerr J, Gurnaney HG, et al. Reducing postoperative nausea and vomiting in pediatric patients undergoing anterior cruciate ligament reconstruction: a quality report. Paediatr Anaesth. 2020;30:446–54.
4. Phillips MR, Adamson WT, McLean SE, Hance L, Lupa MC, Pittenger SL, et al. Implementation of a pediatric enhanced recovery pathway decreases opioid utilization and shortens time to full feeding. J Pediatr Surg. 2020;55:101–5.
5. Echeverria-Villalobos M, Stoicea N, Todeschini AB, Fiorda-Diaz J, Uribe AA, Weaver T, et al. Enhanced recovery after surgery (ERAS): a perspective review of postoperative pain management under ERAS pathways and its role on opioid crisis in the United States. Clin J Pain. 2020;36:219–26.
6. Walker BJ, Long JB, Sathyamoorthy M, Birstler J, Wolf C, Bosenberg AT, et al. Complications in pediatric regional anesthesia: an analysis of more than 100,000 blocks from the Pediatric Regional Anesthesia Network. Anesthesiology. 2018;129:721–32.
7. Guay J, Suresh S, Kopp S. The use of ultrasound guidance for perioperative neuraxial and peripheral nerve blocks in children: a Cochrane review. Anesth Analg. 2017;124:948–58.
8. Ayenew NT, Endalew NS, Agegnehu AF, Bizuneh YB. Prevalence and factors associated with preoperative parental anxiety among parents of children undergoing anesthesia and surgery: a cross-sectional study. Int J Surg Open. 2020;24:18–26.
9. Harbaugh CM, Lee JS, Hu HM, McCabe SE, Voepel-Lewis T, Englesbe MJ, et al. Persistent opioid use among pediatric patients after surgery. Pediatrics. 2018;141(1):e20172439.
10. Cina RA, Ward RC, Basco WT, Taber DJ, Gebregziabher M, McCauley JL, et al. Incidence and patterns of persistent opioid use in children following appendectomy. J Pediatr Surg. 2022;57:912–9.
11. Theroux MC, Lopez M, Olszewsky PJ, DiCindio S, Arai L, Ditro C, et al. Metatropic dysplasia-a skeletal dysplasia with challenging airway and other anesthetic concerns. Paediatr Anaesth. 2017;27:596–603.
12. Theroux MC, DiCindio S, Averill LW, Pizarro C, Oommen A, Bober MB, et al. Tracheal narrowing and its impact on anesthesia care in patients with Morquio A (mucopolysaccharidosis type IVA): an observational study. Anesth Analg. 2023;137:1075–83.
13. Pruszczynski B, Mackenzie WG, Rogers K, White KK. Spinal cord injury after extremity surgery in children with thoracic kyphosis. Clin Orthop Relat Res. 2015;473:3315–20.

14. Drummond JC, Krane EJ, Tomatsu S, Theroux MC, Lee RR. Paraplegia after epidural-general anesthesia in a Morquio patient with moderate thoracic spinal stenosis. Can J Anaesth. 2015;62:45–9.
15. Eiszner JR, Atanda A Jr, Rangavajjula A, Theroux M. A case series of peripheral nerve blocks in pediatrics and young adults with skeletal dysplasia. Paediatr Anaesth. 2016;26:553–6.
16. Alade Y, Tunkel D, Schulze K, McGready J, Jallo G, Ain M, et al. Cross-sectional assessment of pain and physical function in skeletal dysplasia patients. Clin Genet. 2013;84:237–43.
17. Jennings SE, Ditro CP, Bober MB, Mackenzie WG, Rogers KJ, Conway L, et al. Prevalence of mental health conditions and pain in adults with skeletal dysplasia. Qual Life Res. 2019;28:1457–64.
18. Unger S, Ferreira CR, Mortier GR, Ali H, Bertola DR, Calder A, et al. Nosology of genetic skeletal disorders: 2023 revision. Am J Med Genet A. 2023;191:1164–209.
19. Gordon DB. Acute pain assessment tools: let us move beyond simple pain ratings. Curr Opin Anaesthesiol. 2015;28:565–9.
20. Stapelkamp C, Carter B, Gordon J, Watts C. Assessment of acute pain in children: development of evidence-based guidelines. Int J Evid Based Healthc. 2011;9:39–50.
21. Hendriksz CJ, Berger KI, Giugliani R, Harmatz P, Kampmann C, Mackenzie WG, et al. International guidelines for the management and treatment of Morquio A syndrome. Am J Med Genet A. 2015;167a:11–25.
22. Shah SA, Guidry R, Kumar A, White T, King A, Heffernan MJ. Current trends in pediatric spine deformity surgery: multimodal pain management and rapid recovery. Global Spine J. 2020;10:346–52.
23. Wainwright TW, Gill M, McDonald DA, Middleton RG, Reed M, Sahota O, et al. Consensus statement for perioperative care in total hip replacement and total knee replacement surgery: Enhanced Recovery After Surgery (ERAS(®)) Society recommendations. Acta Orthop. 2020;91:3–19.
24. Gustafsson UO, Scott MJ, Hubner M, Nygren J, Demartines N, Francis N, et al. Guidelines for perioperative care in elective colorectal surgery: Enhanced Recovery After Surgery (ERAS(®)) Society recommendations: 2018. World J Surg. 2019;43:659–95.
25. Changoor S, Giakas A, Sacks K, Asma A, Lang RS, Yorgova P, et al. The role of liposomal bupivacaine in multimodal pain management following posterior spinal fusion for adolescent idiopathic scoliosis: faster and farther with less opioids. Spine. 2024;49:e11–6.
26. Rivera-Pérez D, Rosa-Carrasquillo C, Reyes-Sullivan AJ, Torres-Pérez H, Crespo MJ. Anesthetic management including postoperative regional anesthesia in a young adult patient with Von willebrand disease and osteogenesis imperfecta. Cureus. 2023;15:e40363.
27. Pellino TA, Gordon DB, Engelke ZK, Busse KL, Collins MA, Silver CE, et al. Use of nonpharmacologic interventions for pain and anxiety after total hip and total knee arthroplasty. Orthop Nurs. 2005;24:182–90. quiz 91–2
28. Nigam AK, Taylor DM, Valeyeva Z. Non-invasive interactive neurostimulation (InterX™) reduces acute pain in patients following total knee replacement surgery: a randomised, controlled trial. J Orthop Surg Res. 2011;6:45.
29. Ilfeld BM, Gelfand H, Dhanjal S, Hackworth R, Plunkett A, Turan A, et al. Ultrasound-guided percutaneous peripheral nerve stimulation: a pragmatic effectiveness trial of a nonpharmacologic alternative for the treatment of postoperative pain. Pain Med. 2020;21(Suppl 2):S53–61.

Index

M. C. Theroux (ed.), *Anesthesia Care of Patients with Skeletal Dysplasia*,
https://doi.org/10.1007/978-3-031-92501-6

MIX
Papier aus verantwortungsvollen Quellen
Paper from responsible sources
FSC® C105338

If you have any concerns about our products, you can contact us on
ProductSafety@springernature.com

In case Publisher is established outside the EU, the EU authorized representative is:
Springer Nature Customer Service Center GmbH
Europaplatz 3, 69115 Heidelberg, Germany

Printed by Libri Plureos GmbH
in Hamburg, Germany